AT THE END
OF PROPERTY

Patents, Plants and the Crisis
of Propertization

Veit Braun

BRISTOL
UNIVERSITY
PRESS

First published in Great Britain in 2024 by

Bristol University Press
University of Bristol
1–9 Old Park Hill
Bristol
BS2 8BB
UK
t: +44 (0)117 374 6645
e: bup-info@bristol.ac.uk

Details of international sales and distribution partners are available at bristoluniversitypress.co.uk

© Bristol University Press 2024

British Library Cataloguing in Publication Data
A catalogue record for this book is available from the British Library

ISBN 978-1-5292-3366-7 hardcover
ISBN 978-1-5292-3367-4 ePub
ISBN 978-1-5292-3368-1 ePdf

Cover design: Andy Ward
Front cover image: iStock/brainmaster
Bristol University Press use environmentally responsible print partners.
Printed and bound in Great Britain by CPI Group (UK) Ltd, Croydon, CR0 4YY

FSC
www.fsc.org
MIX
Paper | Supporting
responsible forestry
FSC® C013604

In memory of my mother.

Contents

List of Abbreviations

BDP	Bundesverband Deutscher Pflanzenzüchter
CBD	Convention on Biological Diversity
CMS	cytoplasmic male sterility
CPVR	Community Plant Variety Rights
DUS	distinctiveness, uniformity, stability
EC	European Community
EPC	European Patent Convention
EPO	European Patent Office
EU	European Union
GM	genetically modified
GMO	genetically modified organism
IP	intellectual property
NPS	No Patents on Seeds!
PVP	plant variety protection
R&D	research and development
STS	science and technology studies
STV	Saatgut-Treuhandverwaltung
UPOV	Union internationale pour la protection des obtentions végétales

About the Author

Veit Braun is Research Associate at the Institute for Sociology at Johann Wolfgang Goethe University Frankfurt am Main. His research is situated at the intersection of biology, law and economy.

Acknowledgements

Everything that is yours you owe to someone else: this book took quite some time to write and a lot of help from a lot of people. The research presented here was funded by the German Ministry for Education and Research as part of the project 'The Language of Biofacts' (2015–2017), Andrea von Braun Foundation and the German Academic Exchange Service. Thanks go to Suzana Alpsancar, Lukas Breitwieser, Nicole Karafyllis, Uwe Lammers, Sabine Maasen, Barbara Sutter, Franziska Torma, Laura Trachte, Thomas Wieland, Karin Zachmann and especially Johanna Kleinert, who dwelt with me on matters of markets and commodification of plants. Over several years, Saskia Brill, Jeannine-Madeleine Fischer, Oliver Liebig, Laura Kuen and Lisa Rail provided me with intellectual space, fruitful discussions and indispensable feedback. Barbara Brandl passed on her experience to me while Amir Zelinger let me partake in his wisdom; I would have struggled to write this book without either. Birgit Müller, who knows much more about seed than I do, happily shared her knowledge with me. Sarah Earnshaw, Ursula Münster and Paula Ungar always found words of encouragement at the right time. The Multispecies Reading Group and the Agricultural Reading Group at the Rachel Carson Center (RCC) (and everyone who was part of it) helped me to get a better grasp of biology, ecology and agriculture. Thanks also go to Christof Mauch, Katie Ritson, Rob Emmett and everyone else at the RCC who extended their generosity to me. Apologies are due to my students in various seminars (especially 'Property: The Very Idea', spring term 2017) who had to suffer through my attempts to develop the ideas in this book in class.

I cannot consider myself fortunate enough to have encountered Hubert Kempf. Not only did he take me in as an intern even before I had really started this project, but he also opened countless doors for me in the years to follow. This book would not exist without him and everyone else at Secobra, especially Hans Hartl, Seppi Holzapfel, Monika Kratzer, Sabine Rudolphi and Wolfhardt Schmidt, who took the risk of becoming part of a sociological study. Johanna Baron deserves a special thanks for picking me up at the train station every morning. Similarly, Lorenz Hartl and Peter Doleschel (despite his scepticism) at the Landesanstalt für Landwirtschaft

(State Institute for Agriculture; LfL) made it possible for me to get a glimpse of present-day plant science. Theresa Albrecht, Manuel Geyer, Volker Mohler, Michael Müller, Sabine Schmidt, Melanie Stadlmeier, Friedrich Zeller and above all Adalbert Bund shared their knowledge of plants and plant breeding with me, for which I am grateful. Despite their stakes in the debate over patents on plants, a number of people and institutions agreed to interviews. Not all of them have made it into this book, but I could not have written this book without them: Ulrike Amoruso, Alexandra Bönsch and Carl-Stephan Schäfer at the BDP, the Bean Diversity Project at the ÖBZ in Munich, Michael Kock, Christoph Kotschi, David Lahav, Olaf Malek, Ana María Pacón, Wolfram Schiweck, Josef Steiner, Christoph Then, as well as all those who prefer not to be named.

Who knows what would have become of this project if I had not met Hyo Yoon Kang. Not only did she introduce me to a vast literature I had no idea existed, she also took me in at Kent Law School and kept providing me with feedback, knowledge and encouragement. I am also indebted to everyone else in Canterbury, especially Donatella Alessandrini, Steve Crawford, Ahmed Memon, Moritz Neugebauer and Thanas Zartaloudis. Over the years, many people have given me the opportunity to present my ideas and drafts to various audiences, discuss the direction of this project and work out its contents in publications: Sabine Maasen and all the wonderful people at the Munich STS Department; Dagmar Lorenz-Meyer, Josef Barla and the rest of the New Materialism COST action; Kean Birch and Fabian Muniesa; Tilman Reitz; Maria Backhouse and her colleagues from the BioInequalities project; Stephan Lessenich and his group; Alexander Strube at Ackermann Saatzucht; Lukas Fehr, Reinhold Johler and Corrie Eicher; Eva Hemmungs Wirtén and the Passim project; my collaborators Alex Dobeson and Saskia Brill, as well as Phil Roscoe, Liz McFall and Carolyn Hardin at the *Journal of Cultural Economy*; Stefanie Graefe, Daniel Kunze, Lukas Lachenicht, Susanne Lettow, Eduardo Relly, Ute Tellmann and their colleagues from the CRC Structural Change of Property; Jose Bellido and Brad Sherman. I also appreciate Vincent Lepinay's and Sandra Frey's interest in this project, which was encouraging and reassuring during trying times.

This book was researched and written in Munich, but it was re-written in Frankfurt while I was supposed to do other things. I am grateful to Thomas Lemke for allowing me to work on this book; to Katharina Hoppe for pushing me when it was needed; and to everyone else in the Biotechnologies, Nature and Society group who supported me. Special thanks are due to Endre Dányi, Sara Lafuente-Funes and Ruzana Liburkina for helping me stay sane. I appreciate that Doris Schweitzer, Felicitas Sommer and Ute Tellmann disagreed with my interpretation of the law: that I contradict them here means that I did not ignore them. Péter Füzesi encouraged me to think with scripts; the fact that I followed his advice is proof of my indebtedness. My friends and family have supported me throughout the years of this project, and I am glad that they put up with both my unwillingness to explain it at times and my endless rambling about it at others.

Writing a book, let alone publishing it, is not easy. I do not think I could have done it had I not bumped into Paul Stevens. I am forever grateful for his interest, patience, encouragement and support, which carried me through altogether three long, hard years. I would also like to thank everyone else at Bristol University Press for making it possible to turn a very crude manuscript into a real book and assisting me on the way there. The same goes for the four reviewers who read earlier versions of this book and still recommended it for publication. Back when I was struggling with my drafts and ideas, my co-supervisor Helmuth Trischler managed to make me understand my project better every time I consulted him. Maria Gerullis's share in this book is difficult, if not impossible, to overestimate – her curiosity, critique and friendship have allowed me to rethink it many times over. Over the years, no one had to be as patient with me as my supervisor Bernhard Gill. Time and again, I have tested and stretched your patience, Bernhard, but be assured that I am forever grateful for your trust, your honesty and your generosity. I hope I can reward them with this book.

Some of the ideas presented in the following chapters have in part been elaborated elsewhere before in similar or different form:

Chapter 3 builds on V. Braun (2021) 'Holding on to and letting go of seed: Quasi-commodities and the passage of property', *Journal of Cultural Economy* 14: 306–318. https://doi.org/10.1080/17530350.2020.1824934. Chapter 4 systematizes some of the thoughts I developed together with Bernhard Gill in V. Braun and B. Gill (2018) 'Lost in translation: Biofakte auf dem Weg vom Labor ins Patentamt', in B. Gill, F. Torma and K. Zachmann (eds) *Mit Biofakten Leben. Sprache und Materialität von Pflanzen und Lebensmitteln.* Baden-Baden: Nomos, pp 129–154. Chapter 5 expands on ideas first sketched in V. Braun (2021) 'Tools of extraction or means of speculation? Making sense of patents in the bioeconomy', in M. Backhouse, R. Lehmann, K. Lorenzen, M. Lühmann, J. Puder, F. Rodríguez and A. Tittor (eds) *Bioeconomy and Global Inequalities.* Cham: Springer International Publishing, pp 65–84. https://doi.org/10.1007/978-3-030-68944-5_4. Elements of Chapter 6 were originally spelt out in V. Braun (2020) 'From commodity to asset and back again: Property in the capitalism of varieties', in K. Birch and F. Muniesa (eds) *Assetization: Turning Things into Assets in Technoscientific Capitalism.* Cambridge, MA: MIT Press, pp 203–224. https://doi.org/10.7551/mitpress/12075.001.0001.

Where I have borrowed, stolen or misappropriated ideas, concepts and arguments from other thinkers and authors, I have sought to make sure that their names and works are duly cited (unless they wish to remain anonymous). There is little doubt that much in this book has already been said better elsewhere by someone else, and I am grateful that I could rely on such a diverse and profound body of literature as that on property. Whatever original this book contributes to it, I owe to others. The copyright in this work belongs to Bristol University Press; the paper it is printed on is the property of whoever bought this book. Therefore, little remains for me to claim for myself but the errors, mistakes and misunderstandings on the following pages – I am happy to share them with you, dear reader.

1

Introduction

Olaf Malek is not your usual suspect. You would expect a slick business type, complete with pinstripe and an expensive watch, sprung from the script of the latest lawyer series, lecturing you about the importance of intellectual property in a mix of legalese and corporate PR. Instead, the man who has successfully led and won one of the most controversial cases at the European Patent Office (EPO) in recent years has shown up in a button-down shirt tucked into his blue jeans, a bike helmet and trainers. Given how much commotion he has helped cause and how much criticism the case has evoked from across the board, he is surprisingly down to earth. Still, he remains unapologetic. "Nowadays, you might wonder how something like that could have been granted", he leans back and laughs, "but I stand by it."[1]

Malek has every reason to be cheerful. Against public protest, corporate concerns and the opposition of major global players in the food and seed industries, he successfully defended a patent in two rounds before the EPO's highest board of appeal, leading the way for a whole industry. The patented invention itself could hardly be pettier: a tomato that dries on the plant like a raisin when ripening, obtained through mere cross-breeding of wild and cultivated plants and selecting for the desired trait. Yet it is precisely this low-tech nature of the patent which has attracted criticism from an unlikely alliance of activists, plant breeders, legal experts, plant science multinationals and food giants. Despite this broad front and serious legal concerns, the EPO's Enlarged Board of Appeal ruled in 2015 that plants like Malek's tomato are eligible for patent protection. This so-called *Broccoli/Tomatoes II* decision has pushed the boundaries of what is patentable in the European seed industry. For decades, conventionally bred plants occupied a grey zone in European patent law. While many plant bioscience companies and patent lawyers had argued that there was nothing that precluded them from being patented, seed activists, smaller breeders and some legal experts had strongly opposed the patenting of such plants and their traits. *Broccoli/Tomatoes II* firmly put them in the realm of the patentable.

On the surface, the situation seems familiar: in the 1990s and early 2000s, the seed industry was the stage for political controversy (Bonneuil and Thomas, 2009). In what commentators somewhat dramatically termed 'genetic modification wars' (Stone, 2002) or 'seed wars' (Aoki, 2008), activists, seed producers, farmers, politicians and concerned consumers clashed over the promises and risks of transgenic seeds. Disputes were heated and fierce, with activists resorting to the destruction of experimental plots and biotech companies lobbying governments around the world for quick and favourable legislation (Charles, 2001). A key difference between then and now, however, is that most of the seed wars revolved around the benefits and risks of this novel technology for human health and the environment. Although patents on transgenic seed were strongly criticized by activists, it was environmental and consumer protection that they focused on to stop genetically engineered plants. With 'high-tech' out of the picture in the *Broccoli/Tomatoes* case, it is questions of property and its limits that have moved to the centre stage.

The concerns are manifold. Activists worry that 'plants and animals are becoming inventions' (No Patents on Seed!, nd), warning of the privatization of nature. Legal commentators are concerned about the EPO's superficial reasoning in the case, seeing 'legal positivists at work' (Metzger, 2016) at the office. Plant breeders fear for their freedom to operate under a new legal regime, while biotechnology companies are afraid of missing out in a new gold rush. Research institutes and patent applicants worry that the EPO might still reverse course. Meanwhile, farmers wonder what the decision might mean for them; all of this against the background of a wave of mergers and acquisitions among plant science multinationals. In providing a clear answer to a legal issue, the patent office has produced a whole set of new ones.

One possible way of making sense of *Broccoli/Tomatoes II* and the controversy around it is to interpret is as a rupture. Seed, previously off limits to patent law, is being seized, enclosed and privatized, with farmers and consumers missing out. This is the account put forward by some activists, who seek to draw a line between life and nature on one side and intellectual property on the other. In an unprecedented move, and under a flimsy pretence, the EPO and a handful of sneaky patent law firms have overturned both established doctrine and common sense to reap private profits from what should belong to everyone. According to this view, the case has suddenly opened the floodgates for a patenting of life, a realm fundamentally at odds with that of intellectual property law. Some legal scholars publicly put forward a similar view, although their concern is less with profiteering but with a unique misinterpretation of European patent law in *Broccoli/Tomatoes II*, which undermines long-standing principles and doctrines (Metzger, 2016).

An alternative interpretation to this first view is that there is little new about *Broccoli/Tomatoes II*. What is at stake in the case is not an unheard-of category error but an age-old struggle between private property and the commons, between capitalism and the dispossessed, between the haves and the have-nots (Dutfield, 2018). From this point of view, *Broccoli/Tomatoes II* is indeed about private profits at the expense of the public. The difference, however, is that this second interpretation understands the EPO's ruling as just another episode in a long history of privatization in the seed sector, one that the sociologist Jack Kloppenburg (2004) described and analysed in minute detail in his book *First the Seed*. Long before court cases over seed would make headlines, Kloppenburg drew attention to a shift of property rights in seed over the course of the 20th century: from smallholders and public plant science to private research and seed multinationals. Farmers, and by extension consumers, were thus being dispossessed of one of the most essential public goods and the basis for feeding the world, Kloppenburg argued.

This book seeks a third perspective: indeed, *Broccoli/Tomatoes II* is not an unprecedented event, but one of a long series of cases that have redefined our understanding of plants and property. Somewhat contrary to Kloppenburg's long history of dispossession, however, it is not simply the most recent instance of an extension of corporate property rights against public interest. Instead, as I will argue in this book, it signifies a much wider, more fundamental, and multiple crisis of property as such. The fallout of this crisis does not only threaten farmers' interests or people's access to food in developing nations. It also affects those who are commonly thought to benefit from a widening and deepening of property rights: plant breeders, biotechnology companies, patent offices and the prosperous countries of the West. This crisis of property is not limited to issues of distribution, wealth, or access to basic goods: property may be in the wrong hands, but perhaps more alarming, there are increasing signs that it might stop working altogether. Nor is it a crisis unique to plant breeding. Across domains as diverse as digital media, cultural heritage, housing, consumer electronics, or personal data, there is an emerging sense that property is failing. So what is wrong with property, in seed and beyond?

Propertization and its crises

Not too long ago, the picture looked quite different. After the end of the Cold War, it seemed that there were no limits to private property. State companies and assets were privatized in the post-socialist states of the East and the liberal democracies of the West alike, and collectivized property was returned to its 'natural' state, redistributed to former owners and their heirs (Hann, 1998a; von Benda-Beckmann, von Benda-Beckmann and Wiber, 2006a). Not long after, the harmonization of intellectual property rights across the world became a key objective of industry and politics, who pushed for the strengthening of patent laws, copyright and investment protection laws in ever-new free-trade agreements. A third phenomenon was the application of patents to biological matter: suddenly it was possible to think of genes, enzymes and cells as 'inventions', rather than natural substances or body parts (Verdery and Humphrey, 2004). Against this background, critics spoke of a wave of de-appropriation and a 'new enclosure' (May, 2000): what previously used to be in the public domain was now being seized by multinational corporations. There was more going on than just redistribution and privatization, however: the expansion of private property presupposed that the things to be privatized needed to be redefined as something that could be appropriated in the first place.

In many cases, there was no legal precedent to treating, for example, business interests as something that a company owned, and that had to be protected. Similarly, ownership witnessed a redefinition. Franz and Keebet von Benda-Beckmann together with Melanie Wiber (2006a) spoke of the 'changing properties of property', which affected the concept and its definition itself. The UN Declaration of Human Rights, for example, includes the protection of private property, but originally conceptualized owners to be only human individuals. In the years after the Cold War, this was extended to companies, and states that sought to limit or interfere with corporate property rights suddenly found themselves brandished as violators of human rights.[2] It was against this background that Bill Maurer and Gabriele Schwab (2006) saw 'accelerating possession': what was taking place did not just amount to an expansion, it was also an intensification of property rights. For Maurer and Schwab, the Iraq war was emblematic in this respect, signalling a new global order in which new forms of private property would be a cornerstone for the fusion of corporate and state interests.

Hannes Siegrist (2006) offered an alternative view on these events, taking a larger perspective on the role and change of property. For him the expansion and re-invention of property rights around the turn of the millennium did not so much embody a rupture or junction as they continued a historical trend. The whole modern era, Siegrist argued, could in fact be characterized by an expansion and intensification of the institution of property. In this view, neoliberalism, post-socialist privatization or biotechnology patents were only the latest examples of a much deeper historical tendency, a fact that would be obscured by focusing only on the emerging global order after the Cold War. Siegrist termed this longue durée of property-making 'propertization', drawing parallels between the victory march of neoliberal capitalism around the globe and the liberal revolutions of the 18th and 19th century, which fought for private property as the cornerstone of a new bourgeoise society. Over the course of the last three centuries, Siegrist pointed out, property had been applied to a dizzying number of things, from ideas to traditions to body parts; a development that could not be understood by simple references to 'neoliberalism' or 'post-socialism'.

Siegrist's propertization hypothesis may be painted in broad strokes, but it resonates with both our recent experience – the privatization of formerly public assets, the codification of abstract business interests as property titles, the proliferation of intellectual property – and past developments like the historic enclosure of common land in the British Isles (Blomley, 2007) or the creation of modern copyright (Sherman and Bently, 2003). In this view, property emerges as a defining institution not just of a particular period but of the modern era itself. Jack Kloppenburg's (2004) history of plant breeding in the 19th and 20th century can perfectly be read along these lines: as the successive extension of existing property rights and the introduction of new ones that cover aspects and uses of plants hitherto untouched by property. And yet there is a problem with all these perspectives: time has not stopped since the heyday of neoliberalism and globalization; the Iraq War has itself become part of history and can no longer be taken as the symbol for the present. Possession and ownership, too, have moved on, which requires us to revisit what we thought we had understood about property and its logic.

The situation in plant breeding, as outlined, is at the same time reminiscent of and strikingly different from these accounts of propertization, whether they think of it as a short- or a long-term phenomenon. Indeed, the acceleration of possession is still ongoing, as signalled by the European Patent Office's decision to push the boundary of what can be patented. But over a decade after the financial crisis of 2007/08, it is much less clear under which economic and political order this acceleration is taking place. Neoliberalism no longer reigns supreme, raising the question of what is driving propertization today. Even if we take Siegrist's long-term perspective, which stops shy of answering this, we might question if *Broccoli/Tomatoes II* is in fact the latest milestone on propertization's victory march: the broad resistance of the alleged winners of propertization – patent experts, the plant breeding lobby and biotech multinationals – to the ruling should make us wonder if property is still advancing or whether it has started running in circles. This becomes an even more pressing question when trying to read the case through Kloppenburg and his history of expropriation of farmers by seed companies: could it be that this time, it is the latter who are finding themselves on the losing side? If so, where are the winners?

What I am arguing in this book is that while propertization is still continuing, it has run out of steam. It is less clear today than it was 20, 40, or 60 years ago what can and should reasonably be turned into property – and to whose benefit. This means that the critique of property, too, requires reassessment. For the longest time, this critique has drawn into question the economic benefits, the necessity and the naturalization of property. 'Property is theft', the anarchist philosopher Pierre-Joseph Proudhon (1994, p 8) famously wrote, arguing that the exclusive powers granted to an owner of a good – a piece of land, a house, a factory – violated the liberties and legitimate interests of others who relied on it for a living. A similar sentiment can be found in the writings of Jean-Jacques Rousseau, who identifies property as the seed of political order, and the root of inequality. 'The fruits of the earth belong to all and the earth to no one!' he lets his sceptic exclaim (Rousseau, 2011, p 70). Framing property as a matter of inequality and deprivation means to turn it into an issue of political economy: it problematizes access to and distribution of a good, suggesting that this problem can be solved by reforming ownership or getting rid of the exclusive institution of private property altogether, for example by nationalizing or collectivizing it.

As established as this political-economic (or social, as Daniel Loick (2023, pp 43–52) calls it) critique is, it is still popular and widely found in more recent academic work on property. Take Katharina Pistor's (2019) *The Code of Capital*, for example, in which she dissects property law in an international context. Property rights, contracts and other legal tools, Pistor argues, primarily serve the function of accumulating wealth, thus producing inequality. Due to their privileged access to legal and other resources, the wealthy are able to encode their interests in legislation, strategically pick jurisdictions for their financial operations and influence courts in their favour. Thomas Piketty's (2014, pp 377–429) seminal *Capital in the Twenty-First Century* is another point in case. Piketty identifies inheritance as one of the key mechanisms that reproduce and deepen inequality in contemporary societies: protected by the law, wealth accumulated in the past outstrips what a workers could save of their income in the present. Pistor's and Piketty's concerns connect the observations made by Schwab, Maurer and Siegrist with a more systematic critique of property as a vector of inequality and a key tool of modern-day capitalism. To look at property from this perspective means to ask whose financial interests it serves, how it legitimizes and protects them with the help of the law, and how others are losing out as a consequence. Such a critique also gives clear directions for a political engagement with property: first and foremost, we have to identify the winners and losers, the haves and no-haves and alternative arrangements to subsequently change the allocation of wealth.[3]

A second widespread strand of critical thought, which could be called 'moral' or 'ontological' (Loick (2023, pp 53–69) uses the term 'ethical') makes a somewhat different objection to property. It does not so much take issue with property as such but focuses on those instances where it thinks property is wrongly applied (Radin, 1996): living beings, both humans and nonhuman, but also ideas, customs, mountains and rivers and others. Unlike the political-economic critique, this school does not argue that everyone should have their proper share of these. Rather, it claims that the entities in question cannot be property in the first place, as they are not proper objects. It would thus be a category error, and at the same time a moral transgression, to treat them as such. Instead, due tribute must be paid to their 'subjective' aspects, be it their own inherent agency or their persisting ties to human persons. This double argument has a long tradition in the critique of slavery, but it has also been deployed against the asymmetric status of women in marriage, organ trade, the ownership of animals, or against legal conceptions imposed upon people and landscapes by colonial powers. There is also a proximity to a critique of commodification and capitalism here: propertization is viewed as a precondition for turning things into market goods and selling them for money (Hoeyer, 2007). To keep embryos, laboratory animals, blood or genes from being exploited for money, the reasoning goes, they must not be turned into property.[4]

Aside from these two predominant critiques, however, there has recently also emerged a third way of questioning propertization. Perhaps one of the clearest and most intriguing examples can be found in Aaron Perzanowski's and Jason Schultz's (2018) book *The End of Ownership*. How come, they ask, we increasingly end up buying things without ever becoming their owners? Music, books, video games, even coffee machines, cars and printers, Perzanowski and Schutz point out, are no longer 'ours' the way they used to be: long after we purchased them, the sellers can change their minds about these goods, add or remove features, prohibit us from repairing them, dictate how we use them or revoke access altogether. The reason is the increasing importance of intellectual property rights, which have found their way into our homes in the wake of digitization and continue to govern the use of what should rightfully belong to us alone, they argue.

If the critiques as described earlier are 'political-economic' or 'moral-ontological' in nature, we might call Perzanowski's and Schultz's 'cultural' or 'cultural-economic'.[5] What is at the heart of their critique is not an idea of property as a vector of inequality and misallocation: after all, no one is going to starve if Amazon deletes the books they bought from their e-reader or Tesla remotely changes how fast their sports car can drive.[6] Neither is it an indictment against property misapplied, as both think that the things they examine *should* in fact be property – they simply have been propertized the wrong way. The authors' criticism is instead aimed at the demise of a lifestyle, a loss of sovereignty over our property, a helplessness in the face of after-the-fact changes to the terms of ownership that we agreed upon at sale.

Instead of enjoying the freedoms of consumption, consumers witness corporate legal departments intruding into their personal sphere, barring them from what until recently they were used to do. What Perzanowski and Schultz call for, consequently, is not a revolution in which private property is seized and redistributed, or the liberation of unruly things from their property shackles. Quite to the contrary – they ask for nothing more than the reinstatement of 'ownership' as it existed before the rise of digital media and the infusion of technical objects with software and code. Rather than a fundamental critique of property, this book makes a modest but determined defence of it. Private property here is a way of life and an integral part of liberal consumer culture. *The End of Ownership* thus mobilizes one form of private property – personal ownership – against another one – intellectual property rights.

This is not to say that the political-economic or moral-ontological critique of property have fallen out of time, nor that a cultural critique is superior to them. What Perzanowski's and Schultz's commentary merely highlights is that more is at stake. After all, there is a long tradition of thought that has embraced property not as a tool of accumulation, exclusion, commodification and profiteering, but a means of autonomy and liberation: from the agrarian land reform movements of the 19th and 20th century (Kautsky, 1988; Dobeson and Kohl, 2023) to Virginia Woolf's (1929) *A Room of One's Own*, to socialists' ideas about collective ownership and management of farms and factories, to a feminist movement that claimed property for women in their own bodies (Pollack Petchesky, 1995; Dickenson, 2017), propertization has also taken the form of an emancipatory project. These *values* we attach to property are often overshadowed by a singular emphasis that the political-economic critique puts on *value* and the objectification and oppression the moral-ontological critique sees in propertization. Therefore, rather than naturalizing property as an object of value like the political-economic tradition tends to do or deconstructing it through a moral-ontological critique, the aim here is to *reconstruct* it. Property has certainly been a vector of inequality and objectification, but in many ways, it has also been a vector of progress and emancipation (Alexander, 1999). For all our reservations about property, private ownership and oppression, we are still wedded to the idea that there is also *good* property, be it private or collective.

With this book I seek to show that laudable as this idea is, it is becoming increasingly infeasible. As much as we wish to return to the good days of old when our music albums and books were 'really' ours, when property was undivided and unequivocal, I do not believe this is going to happen. To make things even worse, it is not only consumers who are on the losing end. As this book will show, the alleged beneficiaries of propertization – companies large and small – are increasingly troubled by property issues, with no remedy in sight. Ultimately this, too, is a political problem: one that is not easily captured or solved by the political economy of distribution but one that requires debate and action nonetheless. How should we deal with the erosion of property as we once knew it? If propertization has indeed started running in circles, does that mean the end of a progressive project? By delving deeper into the confusing state of property in plant breeding, I will try to answer these questions and shed light on 'property condition' of our times.

Plants and property

Why look at plants and seeds, of all things, to inquire into the nature and state of propertization? The choice is both obvious and requires explanation. Obvious because there is a long tradition of works that have used plant breeding to think about property: next to Kloppenburg's (2004) *First the Seed*, authors like Deborah Fitzgerald (1990), Cary Fowler (1994; 2000), Daniel Charles (2001), Thom van Dooren (2007a; 2007b; 2008), Alain Pottage and Brad Sherman (Sherman, 2008; Pottage and Sherman, 2007; 2010; 2011), Jay Sanderson (2017) and many others (such as Hayden, 2003; Müller, 2006; 2015; Schubert, Böschen and Gill, 2011; Harwood, 2012; Brandl and Glenna, 2017; Chapman, 2018; 2022; Girard and Frison, 2018) have used plants to discuss property and vice versa. There is thus a rich literature on plants and property to draw from – which of course also begs the question of what could still be added to it.

The reason for this continuing interest in the field certainly owes to the abundance of overlapping and diverging property regimes in the seed industry. For over three decades now, plant breeding has been at the frontier of propertization, witnessing a staggering proliferation of ever-new forms of property which have become increasingly controversial. As a result, one and the same plant could in theory be subject to over half a dozen different property regimes, among them physical property, plant variety protection, patent rights, hybrid seed, trade secrets, terminator technology, protected denomination of origin and resowing remuneration and an according number of different proprietors.[7] What intellectual property lawyers call a 'patent thicket' (Frison and van Zimmeren, 2021) – the sprawl of overlapping and contradictory patent claims in a particular technology – has become a wider condition here, evidenced by the *Broccoli/Tomatoes II* ruling: we are not only dealing with problems in patent law, but in many other property regimes as well, as we shall see throughout this book.

There are two predominant ways of telling the story of property in seeds. One follows the political-economic critique of property, seeking to show how the history of propertization in plant breeding is a history of expropriation: in this story, seed is taken away from farmers into the hands of plant breeding companies (Fitzgerald, 1990; Kloppenburg, 2004; Harwood, 2012). The latter then use their strengthened property rights to squeeze more profits from the former and, by extension, from consumers who depend on the food grown from seeds. A second, somewhat different story is that of the misapplication of intellectual property, especially patents, to plants (Pottage and Sherman, 2011). It mainly arose from discussions and legal disputes over the use of genetically engineered seed. The case of the Canadian farmer Percy Schmeiser, who was sued by the seed company

Monsanto over the unlicenced reuse of their patented rapeseed, was especially influential in that regard (Charles, 2001; Braun and Gill, 2018).

In the wake of this and similar cases, commentators asked whether the patenting of plants was not a grave category error. After all, plants were neither invented nor intangible ideas, and many of the core tenets of patent law were led ad absurdum as a consequence. While both narratives overlap, the underlying assumptions about property differ: in the first case, property is often presented as an unambiguous, well-working instrument that serves the interest of seed companies, whereas in the second case, the grammar and logic of property are problematized, without necessarily commenting on the economic consequences. Both of these discourses are again mirrored in activist strategies and political debates, which either ask for the economic effects of propertization or draw into question the applicability of property to plants as living beings or natural objects.

Given these extensive debates and the high level of politization in the seed wars, it is little surprising that plant breeding has become such a fertile ground for the study of property. What is puzzling, however, is why this happened only very late: the first edition of Kloppenburg's *First the Seed* appeared in 1988, before genetically engineered had caught public attention (Charles, 2001), and within the span of a few years, a number of important books and studies followed (Fitzgerald, 1990; Fowler, 1994; Charles, 2001). By the early 1990s, intellectual property in plants had already been around for a while, however: in 1930, the US Congress passed the Plant Patent Act, which created a special patent for certain crops. After World War II, Western European countries introduced plant variety protection (PVP), a property regime for plant varieties. Yet it took another 30 years before property in plants became widely politicized, raising the question of what changed between then and now. Indeed, as I want to show in this book, both the politics of propertization and the space it operates in have changed over time. The plant breeding Kloppenburg was writing about in the late 1980s was different from that of the early 20th century, and since then the industry has further changed.

It is not just intellectual property that makes plant breeding an interesting case, however. Plants are situated at the boundary of immobile land and mobile possessions, of living being and technology, of object and subject. Most of the literature has focused on the paradoxes and contradictions that emerge when patents and other intellectual property are applied to plants and seeds. The unwritten assumption behind this approach is that because intellectual property in plants is flawed concept, physical property in plants is not – and thus by default the 'proper' way of owning plants. There are good arguments for this. After all, plants are not ideas, inventions, texts or codes, but tangible, physical beings. But this also means that they are not simply inanimate objects either (van Dooren, 2007a), which is a conceptual challenge: most theories of property, both emphatic and critical, rely on the assumption that the object of property is mute and inactive, while the real agency flows from a subject.

It is tempting to simply claim the opposite – plants are in fact subjects (Stone, 2010)! But that claim is equally difficult to defend, which leaves us with the question of what role plants should ideally take in property arrangements. Much of the difficulty with plants already starts with defining their physical boundaries: when I buy a seed, is the grown plant contained in it, too? And what about the seeds it will produce? These have become contentious questions in the last 30 years (Braun, 2020), but they were difficult to answer even before that.[8] To complicate things further, in the wake of advances in biotechnology and molecular biology, it makes much more sense today to understand plants as technology, DNA and genetic code (Calvert and Joly, 2011), even if this understanding has remained incomplete (Fox Keller, 2005).

More than law: on the nature of property

Many of the works that study property start with the law, expecting that, ultimately, it will tell them what property is and what it is not. It is then up to social scientists, economists, philosophers and so on, to spell out the consequences of law's definition of property, but this remains a one-way street: the law belongs to legal scholars, and if we want to critique property in detail or essence, we will have to be (or become) legal experts.[9] As a result, the critical study of property is all too often 'property law plus' – property law plus economics, property law and society, property law and ethics and so on. The popular notion of property *rights* (Carruthers and Ariovich, 2004) is already evidence of this tendency: it assumes that property consists of legal rights, which confer and constitute property.[10]

It is problematic to understand property this way. Is there no property before the law? To my surprise, whenever I have raised this question over the past years, few colleagues have been inclined to answer with 'yes, there is'. Only through law, they have argued, are the terms of property spelt out at the same time and created, hence why any inquiry must start and end with them. Other forms of ownership, belonging and possession, it is further argued, simply do not qualify for treatment as property, and should accordingly be analysed in completely different terms. While I can understand this reasoning, I object to the matter of course with which it cedes property to law.[11] After all, legal definitions for things like 'marriage', 'corporation', 'war' or 'money' have stopped few scholars from other fields and disciplines from coming up with definitions of their own, many of which make do without any legal elements. To understand property as 'made from law' would mean to understand propertization and its crises as an ultimately legal question, when in fact it is much more than that.

In this book, I will thus pursue a concept of property that does not rely on a privileged role of legal definitions, rules and texts. The fact that law is often the most visible or easily accessible element of property only means that it is the tip of the iceberg: below the surface, much more is going on. If I have previously argued that propertization is a cultural, not just economic or moral condition, I also want to stress that property is not primarily a legal but a more general, cultural phenomenon. There are many ways to make property work, I will try to show in this book, and law is only one of them. While it indeed makes a considerable contribution to some forms of property, it is negligible or absent in others. Just like it is not always the law that takes away your things from you, it is not necessarily what will return them to you.

Bicycles are a good example. Imagine your bike is stolen: what can the law do to return it to you, or to prevent the theft in the first place? The police, despite acting 'in the name of the law', will certainly not try to get your hopes up when you show up to report it, nor will thieves be particularly impressed by your legal status as the rightful owner (Johnson et al, 2008; Whipple, 2023). Your best chance is if you have properly locked your bike or put a tracking device on its frame, which might allow you to retrieve it (Pratelli et al, 2017). On the other hand, a considerable number of bikes end up 'orphaned' at lost-property offices. Ample time and money are spent maintaining these curious institutions whose task it is to rejoin owners and their property.[12] Yet despite the cooperation of honest finders, and to the frustration of the offices' staff, their attempts all too often remain unsuccessful (FitzGerald, 1895): the problem is not that the law is unclear but that the rightful owner cannot be identified – or does not care in the first place.

As we will see in this book, this 'more-than-legal' nature of property is difficult to overlook in plant breeding. In producing seed, plants violate patents but cannot be held liable (Bernhardt, 2005); hybrid genomes and terminator technologies step in where legal texts alone cannot keep farmers from misappropriating plants (van Dooren, 2007b); states pass laws that prohibit multiplication and sale of seed but see themselves unfit to enforce them (Braun, 2020). Perhaps just as important, there is a practical and material dimension to property in plant breeding that would be gravely neglected by framing it as a legal phenomenon, or by only asking about just access and misapplication of property status. How, for example, can farmers be sure that they actually make seed they buy 'theirs' on the farm? How do companies get hold of resistance genes to incorporate in the genomes of their plant varieties? How are contracts over seeds and harvests honoured in practice by the parties who have agreed upon them? These are questions that legal analysis alone cannot answer. Property in plant breeding (like elsewhere) is as much about practices and technologies as it is about laws, rules and rights.[13]

This more-than-legal nature of property is both trivial – we are all familiar with bike locks, name tags and door keys – and a challenge. There is an overwhelming amount of literature on property as rights, but comparatively little has been written about property as practices, and even less on property as technology. The difficulty lies in joining the various existing strands of property thought while expanding and connecting the fragmentary ideas about the materiality of property in a way that allows them to speak to each other (rather than mobilizing them as each other's critiques). My own attempts in this book are not meant as an answer to what a more-than-legal concept of property *should* be, but as a proposition for what it *could* look like. What I hope to show through the example of plants is how fruitful a widening of established concepts and a thinking outside of disciplinary, theoretical and methodological boxes can be for the study of property.

Theory and method: a few notes on genre

But if it is not confined to legal texts, how to study property? Throughout this project, I have been unsure how to answer this question. It hinges on discipline: what are proper questions to ask about property, and what not? Who is the audience for the answers you obtain (Fleck 1981, pp 38–51)? There is no overarching theory that would allow us to speak about property as such but rather a few dozen schools of thought that engage with legal, philosophical, economic, anthropological or political aspects of property; each from its own vantage point. My own training is in sociology, a discipline with a mixed record when it comes to the study of property. Landed property was an important topic at the beginnings of sociology in the 19th and early 20th century (Tönnies, 1926; Weber, 1950; 1978; Kautsky, 1988). With the demise of rural society in Western countries, however, it faded from the discipline's consciousness until the mid-20th century (Dobeson and Kohl, 2023). After World War II, sociologists little systematic attention to the topic, with the exception of a few (unsuccessful) attempts to revive the sociology of property (Gouldner, 1970; Swedberg, 2003a; Carruthers and Ariovich, 2004; Gill et al, 2012). As a result, sociology has relatively little to say about property today,[14] especially when compared to social and cultural anthropology.

The anthropological study of property reaches back to the 19th century, too, but is somewhat more continuous. It benefits from its comparative approach and study cases in which the state and its law cannot be relied upon to understand explain property. One of its earliest and most important contributions, as I explain in more detail in Chapter 2, is that of property as a bundle of rights and a social relation between humans. Another one is its insistence on the importance of practices and their close relationship with more or less informal rules, rather than codified legal texts (Turner, 2017).

Its 'property renaissance' in the 1990s and early 2000s, driven among others by Marylin Strathern (1988; 1999), Chris Hann (1998; 2007), Franz and Kebeet von Benda-Beckmann (von Benda-Beckmann, 2006a), as well as Melanie Wiber and Bertram Turner (Wiber, 2015; Turner, 2017; Turner and Wiber 2022), can also be regarded as both more profound and lasting as the short-lived revivals in sociology. The anthropology of property has provided important impulses for other fields, such as ecological economics (Schlager and Ostrom, 1992) and also was a key resource for me when writing this book.

Nevertheless, I have found that anthropology's widespread insistence on practice and its comparative outlook can also be a disadvantage at times. They are sometimes mobilized as counter-concepts to an allegedly hegemonic or monolithic 'Western' property that relies on laws and the state and consequently remain confined to the messy frontiers of the Western legal order (Hann, 2006; Strathern, 2004). Practice and social relations, like law, further locate the source of property in the human element in property relations. This is an assumption difficult to uphold in the face of plants' very obvious contribution to the working and failure of property arrangements – or only at the price of rendering it invisible. If genes, seeds and harvesters are neither laws nor practices nor relations, how do we account for their agency?

Many of my own attempts to answer this question owe to work done in the field of science and technology studies (STS), in which researchers from various disciplines have sought to understand science and technology as collective efforts of a social nature, rather than realms completely separated from the rest of society (Biagioli, 1999). From the 1970s on, ethnography in scientific laboratories has played an important role in STS as a way to uncover 'science in action' (Latour, 1987), that is, practices, controversies and ambiguities, rather than a consolidated textbook account of science as straightforward, rational and progressive (Doing, 2008). Somewhat in contrast to classical sociology and anthropology, STS has also developed a strong stance on the active role of 'non-humans', both in laboratories and beyond – in markets (Callon, 1998a; 1999), parliaments (Danyi, 2018), courts (Latour 2010) and so on.[15] We cannot properly understand science, technology or society, STS argues, unless we pay due attention to what objects, infrastructures, organisms and devices contribute to them. It is not just for the discussion of the science of plant breeding, therefore, that this book strongly relies on this tradition but also for highlighting the material dimension of property.

There is unfortunately little discussion of property in STS (even less than in sociology),[16] and its understanding of law (like that of most social sciences) remains very rudimentary.[17] Much of what is written in the following chapters therefore draws from a neighbouring field, which combines critical legal studies, the history of science, media studies and STS alike. While to my knowledge it does not bear a proper name, it could well be called 'history and theory of intellectual property', as many of its protagonists congregate in the *International Society for the History and Theory of Intellectual Property* (ISHTIP). Despite its focus on intellectual property, this school is very much interested in the materiality of legal texts and allegedly 'immaterial' property objects as well as the logic sedimented in intellectual property law. Although a comparatively small and young field, it has produced a remarkable number of important works (Sherman and Bently 2003; Biagioli, 2006; Pottage and Sherman, 2010; Biagioli et al, 2011; Kang, 2019; Kang and Kendall, 2020), and its engagement with plant breeding is extensive (Kevles 2007; Sherman, 2008; Pottage, 2011; Pottage and Sherman 2011; Sanderson, 2017; Bellido and Sherman, 2023). These works have been indispensable for writing this book, and while I cannot claim that it follows in the same tradition, it shares many of their conceptual intuitions.

Such a crude mix of intellectual resources is already a problem in theory, since the approaches mentioned, despite their commonalities and overlaps, eventually diverge. Perhaps more importantly, they are a problem in practice, too: what research method could claim to respond to them and to capture the full complexity of property of plant breeding? Following the example of STS laboratory studies, I initially tried to write this book as an ethnography. I quickly had to learn that this was unsatisfactory and incomplete. The development of a plant variety spans ten years and more, with several seasonal cycles of sowing, evaluation and harvesting in between, a process I could not possibly have covered even if I had written this book twice as slow as I did. Another thing I had to learn is that ethnography, owing to its roots in anthropology, relies on a set of assumptions – about distance, separation, strangeness, hierarchy – which may not always be given in a research context. My first field sites were barely two hours from my home, and it turned out that 'science in the making' (Latour, 1987, p 4) in laboratories can be less important and insightful than 'ready made science' (Latour, 1987, p 4) in patents and textbooks. I soon found it difficult to separate research subjects from colleagues or to turn my lack of biology knowledge into a resource when what was needed was a grasp of fine details. There is a long discussion about 'studying up' in ethnographic literature (Nader, 1972; Gusterson, 1997; Plesner, 2011), which problematizes the quiet assumption that participant observations and interviews will be conducted from a position of power and asks what happens when tables are turned. Over the course of my exchanges with plant breeders, patent lawyers, business leaders and activists, I have been both on the long and the short end of hierarchies, but in many cases, this was not what made discussions productive or difficult.

What is more is that plants, patents and even some institutions do not easily lend themselves to being studied ethnographically. Seeds do not speak, unless they are either cut up and analysed or planted and grown. Perhaps there are ways to understand the language of patents through participant observation, but unlike others (Parthasarathy, 2017), I did not manage to find my way into a patent law firm or even the European Patent Office. Some things, like numbers and histories, are notoriously difficult to capture through oral methods. For these and other reasons, I have pursued other methodological approaches to them: document analysis, the use of databases for both statistics and as archives and expert interviews (Bogner, Littig and Menz, 2009) to complement them. In interviewing experts, it was important for me to speak to people representative for the various factions in the *Broccoli/Tomatoes II* controversy: seed companies large and small, plant scientists, patent lawyers, anti-patent activists, lobbyists and growers.[18] Most of the research was conducted during my PhD studies at LMU Munich from 2015 to 2018, coinciding with the peak of the *Broccoli/Tomatoes II* debate, but I have also followed up its aftermath and stayed in touch with the people involved in it in the following years, up to the finalization of this work. The research for this book was to the largest part carried out in Germany. Some interviewees were also located elsewhere: the Netherlands, Switzerland and Israel.

The result is a book that tells a particular story: that of plant breeding, situated in Europe, notably Germany, largely from the perspective of seed companies and patent lawyers. But I hope it is also a book that extends beyond these contexts; one that sheds light on the complex nature of property, on its proliferation in so many different domains, and on the problems that are caused, rather than solved, by the creation of ever-new forms of property. In the chapters to follow, I want to open up property: as a concept and an object of study for various disciplines, but also as a question that is posing itself in ways that cannot easily be answered anymore.

Outline of this book

The first opening in Chapter 2 is a theoretical one: what could a more-than-legal concept of property look like? The difficulty is to remain sensitive to the differences between legal, practical and other elements that contribute to property while studying how they work together. The chapter starts from the established legal notion of property as a 'bundle of rights' (Penner, 1996; Orsi, 2021), which carries a sense of diversity and composition. This composite understanding of property has some major advantages, such as its idea of multiple, overlapping and thus relational ownership, the legal fragmentation of property objects, or a 'division of labour' between rights of different types in producing property. Its great disadvantage, however, is that it still thinks of property in legal terms. Borrowing from anthropology and institutional economics (Schlager and Ostrom, 1992; Bourdieu, 2008), I thus make a case for a corresponding 'bundle of practices', as property rights usually pertain to particular activities, process and uses.

To speak of property as a set of practices, rather than rights, moves social activities (rather than just relations) to the centre stage. The flip-side, of course, is that it comes with two analytical distortions. The first is an opposition, rather than collaboration, between rights and practices; between property prescribed and property performed. The second is a blind spot for those elements that are neither law nor practice and would hence be overlooked in framing property as either one or the other. In a final step, the chapter therefore expands the bundle again: from practices to scripts, which carry the threefold meaning of (legal) texts, dramaturgical plots and technical logics. Rather than answer what property is made of, the 'bundle of scripts' asks how different elements need to work together to produce effects like exclusion, appropriation or alienation.

Two of these scripts are prominently featured in Chapter 3, which provides an introduction to the plant breeding business, its practices and its history. Modern seed does not come in all shapes and sizes, but in the form of standardized varieties. It is a peculiar object: not just because it is the product of a particular technological process (Gill, Torma and Zachmann, 2018), but also because it is traded in transregional markets – a fact that represents a challenge for breeders and farmers alike. In market transactions, goods are alienated and appropriated at the same time – the buyer acquires what the seller lets go of (Slater, 2002). But how do the parties involved make sure that neither too much nor too little changes hands? This is not a merely theoretical concern but one that occupied the fledgling seed industries on both sides of the Atlantic well into the 20th century. The appearance of commercial seed opened the question of what belonged to farmers and what to breeders, two professions that were only beginning to separate. What I argue in this chapter is that in the mid-20th century, this economic and sociological problem could still be answered with *more* property: with the introduction of PVP legislation in post-World War II Western Europe, lobbyists and legislators found a solution that created a working seed market by undermining the market principle of perfect alienation and appropriation. To this end, seed had to be augmented with a set of scripts that allowed breeders to retain the exclusive ability to commercialize seed, while farmers could otherwise use it as they saw fit, creating the illusion of sovereign, exclusive property on both sides.

Elegant as it was at the time, the PVP framework left a number of questions unanswered that would become more pressing in the decades to follow. While the European solution successfully redistributed property between breeders and farmers, it relied on a tacit collaboration between seed companies when it came to acquiring new plant traits and appropriating them to the collective pool of plant varieties. Over the course of the 20th century, the production, appropriation and distribution of traits became more and more important and called for property regimes of its own, a history that is reconstructed in

Chapter 4. While patent protection for plant traits was discussed early on in both the United States and Europe, it was rejected in favour of bespoke property regimes, as the biology of plants and the practices of breeding were deemed incompatible with the logic of patent law (Fowler, 2000; Pottage and Sherman, 2011; Sanderson, 2017).

With the advent of biotechnologies like DNA sequencing and genetic transformation, however, plants changed both materially and conceptually: they could suddenly be understood as emerging from laboratories instead of nurseries, as textual and chemical machines rather than living organisms and as consisting of genes instead of forming a variety. This redefinition of plants allowed patent law to extend into the domain of plant breeding, which had previously been regarded as off limits. Even after the marginalization of many of the novel technologies in plant breeding practices, especially after the public controversies over GMOs (Winickoff et al, 2005), the logic of patent law continued to expand, from transgenic to so-called 'native' traits. The remaking of plants in the second half of the 20th century thus laid the foundation for what would ultimately culminate in *Broccoli/Tomatoes II*: the question of how much property plants should and could accommodate without threatening the future of the seed industry.

Why exactly are patents on plants so contentious? There are two predominant views in economics on patents and what they do for businesses and industries (Machlup, 1958). The first, claiming a positive role for them, holds that without patents, firms cannot innovate, as there would be no instrument for them to recoup their investments: competitors would simply freeride on an innovators' R&D expenses (Callon, 1994; 2021; Landes and Posner, 2003). The second, sceptical one argues that the transaction costs of patents – application fees, licensing negotiations, royalties – often outweigh their benefits and in fact slow down the rate of innovation and technology development (Shapiro, 2000). Both views, however, frame intellectual property as an economic tool whose purpose lies in channelling revenues.[19] Speaking with various stakeholders in the *Broccoli/Tomatoes II* controversy however reveals that it is very difficult to tell what patent actually do in and for the seed sector.

Anti-patent activists fear they will lead to market concentration and oligopolies, eventually squeezing more money out of farmers relying on seed. Some plant breeding firms fiercely object to patents on seeds yet keep heaping up patents. Others endorse them while drastically cutting their portfolios. Some scholars and breeders argue that patents have become empty signifiers that only signal innovation to shareholders and stock markets without ever producing it (Kang, 2015). Others point out that patents have become empty signifiers within companies' bureaucracies themselves: firms keep filing patents without knowing what they do. In dissecting the competing and contradictory theories about what they do in practice, Chapter 5 shows how patents, far from embodying an economic or technological rationality, are cultural phenomena: what they do (or do not do) depends on company culture as much as on what others think a company's culture is. What most voices in the seed industry agree on is that what is needed is less, rather than more patents – and yet, taken together, their different patent cultures only help patents proliferate.

If discussions around *Broccoli/Tomatoes II* have predominantly portrayed patents as a problem, they largely assume that PVP and other forms of property work well. In Chapter 6, I want to show that this is a simplistic view of things. Patents are not the only problematic property regime at work in plant breeding – PVP, too, has become less and less sustainable since its inception in the mid-20th century. Unlike 50 years ago, plant breeders no longer feel the post-war consensus in the seed sector can support their businesses (Sanderson, 2017). While they successfully lobbied for amendments to the original PVP laws in the 1990s entitling them to royalties for seed reuse, these revisions have been only partially enforceable: the state as well as other players in the plant breeding industry are reluctant to help breeders claim 'their' property against farmers, who continue to resow varieties without remuneration.

Left hanging by the law, breeders have sought to reclaim their property by other means. The use of hybrid breeding systems promises to deliver plant varieties that cannot be reproduced true to kind and that will force farmers to buy fresh seed every season (Fitzgerald, 1990). The viability of such systems is also an economic question, however: given the low yield advantages of hybrid seed and high level of competitiveness in many agricultural crops, farmers and seed producers are unlikely to switch to hybrid varieties anytime soon. To make matters worse, plant breeders are now also finding themselves at the receiving end of propertization. The wild plant races and genetic resources they have long relied on for improving their varieties have become the 'national property' of the countries of origin and local communities under the Nagoya Protocol on Access and Benefit-sharing. Suddenly, breeders need to find ways around the restrictions imposed by this new property regime and have to realize that what they thought was theirs now belongs to someone else. This turns the property relations Kloppenburg (2004) described and criticized upside down: the political compass we have long used for judging the politics of property cannot guide us here.

The final chapter returns to the questions posed in this introduction. Why has property run out of steam? Why do people like Olaf Malek, despite acknowledging the shortfalls of property in plant breeding, welcome its proliferation or even ask for more of it? Why does the seed industry seem incapable of thinking beyond property? And if property is no longer a solution, what is? There was a time, I argue, where property indeed helped solve social issues. The ill-functioning market for seed is certainly the best example, and it is little surprising that the memory of this success is still alive in the industry. But such solutions were never built on property alone: they relied on a certain idea about what a breeder and a farmer, a producer and a consumer, an organism and an invention, a subject and an object was. If they worked, it was because there was still space for property to organize, but also enough social distance between the people who all saw themselves as owners, allowing them to overlook the fact that the things that belonged to them also belonged to others.

The more property expands and the more we abandon our roles as farmers, consumers, competitors, innovating businesses or subalterns, the less we can overlook that our property is manifold and entangled (Thomas, 1991). The 'Lockean topology' (Pottage, 2004) that propertization promised to impose on the modern world is showing its cracks, and more property will not be able to fill them. Proposals in the seed industry for where to go from here range from forging trust and personal relations rather than new property to cutting down the bundle of property to a bare minimum. And yet it is difficult to part with property after all it has done for plant breeding: after all, it is one's own property that the sector will have to forgo. This is a problem that extends beyond the seed sector: economically, the old property of full ownership looks increasingly unviable, as do political visions that promise fair distribution or property for all. The crises of property can be felt in the music business and the software industry, in publishing and in post-colonial debates alike. But if we have indeed reached the end of property, can we let go of it?

2

From Rights to Scripts

How to sort out property? If you inherit, want to sell a house or seek to rent a piece of land, this is very much a practical question. Answering it all too often entails engaging with statutes and fine prints, consulting legal experts and finding a path through the thicket of property law. Although we often have a clear sense of what is, or at least should be ours, the law does not necessarily agree – and making our own ideas about property match those of the law can mean a lot of legal work. Sorting out property, however, is also a theoretical concern. When writing about property as scholars we, too, all too often assume that we first and foremost have to turn to the law to find answers.

'Those ... who make use of the words property ... before they have explain'd the origin of justice ... are guilty of a very gross fallacy, and can never reason upon any solid foundation', David Hume (1960, p 491) writes in his *Treatise of Human Nature*. After all, Hume argues, this is what property is built from and on: 'Our property is nothing but those goods, whose constant possession is establish'd by the laws of society; that is, by the laws of justice' (Hume, 1960, p 491). Therefore, to speak about property, we must first indulge in the law, Hume demands. Countless volumes have since followed this demand, to the effect that today we understand property first and foremost as a legal phenomenon. Whether it is the discussion of property in the constitutional order of the republic (Ryan, 1987; Macpherson, 1990; Alexander, 1999), the debate over the effects of property law on inequality (Beckert, 2008; Errygers and Cunliffe, 2013; Piketty, 2014) or the attempts to sketch out a sociology of property rights (Gouldner, 1970; Carruthers and Ariovich, 2004), all these works assume a fundamental or privileged role of the law in making property.

There are good reasons for thinking about property as legal. One is that legal processes produce an endless number of texts, which offer the benefits of close examination, easy citation and at least superficial stability in space and time.[1] Another is comparability: no matter how different or similar we believe societies across the globe to be, we somehow always assume that they have institutions and rules that can be called 'law' (Latour, 2013, p 371). It is for this reason that anthropologists like Marilyn Strathern (1988) can juxtapose Melanesian with Western property rights, or that we can expect to gain insights from Max Weber's (2008) historical discussion of ancient agrarian law. To understand property as a legal issue further gives us clear directions for tackling the problems tied to it: to fix property, we need to change the law. The legal also serves as a useful contrasting foil for our critique: we can point out all the political, economic and social factors that go into forging it (Feenan, 2013; Parthasarathy, 2017; Pistor, 2019), thus undermining the law's air of autonomy and neutrality as well as the privileged status of legal experts.

At the same time, however, all these things are also disadvantages. The study of legal texts does not tell us how we move from the letter of the law to that which it is supposed to govern – in this case, living organisms. Can the law, by itself, bring property into being? Or does property law rely on other elements to work? Can we perhaps even imagine property without the law? Some of these questions have been troubling the comparative perspective in those cases in which it had to deal with unwritten, sometimes even unspoken rules (Bourdieu, 2008). Is it still plausible to think of customs and practices as 'law' even if things are simply 'done' rather than rules made explicit and carried out (von Benda-Beckmann et al, 2006b; Wiber, 2015)? If property relies on habits and practices rather than explicit, codified rule, how would one go about to change it? The critical move of pointing out that the law is mixed with other logics quickly becomes pointless here, as it can be difficult to distinguish them in the first place.[2]

These are more than just theoretical concerns. The problems that we are witnessing in the wake of propertization are not simply located in property law or the legal realm. They also involve the material nature of property objects as well as the practices that unfold around and between them. In order to address what is wrong with property today, we need to go beyond the confines of the legal and learn to grasp property as a more complex arrangement of legal, practical, technical and other elements. What I will do in this chapter is sketch out one way of doing this: by moving from the popular concept of property as a 'bundle of rights' to a more comprehensive bundle that can account for all the non-legal ways in which property can be made, sustained and unmade. In broadening the concept of property this way, we will not only be able to take into account its heterogeneity. Property as a bundle of scripts will also allow us to understand better how and where it is forged and maintained, rather than merely decreed: on fields and in laboratories, in spreadsheets and in biology textbooks, in genomes and in business relations. Finally, it will also give us a sense of property's own agency, one that all too often exceeds and undermines the intentions of owners and property law – and one that is becoming increasingly problematic, contradictory and difficult to contain.

The arithmetic of property: one, two or many?

There is a profound disagreement between everyday language and scholarly accounts when it comes to the meaning of property. Take one of the most obvious examples, the use of the term 'property' in English for plots of land and buildings. 'This property' or 'a property' refers to a given, more or less clearly demarcated object in space (for which more technical texts reserve the term 'real property'). The basic unit for property here is therefore one: *a* property. We could call this model of property 'monadic' in a double sense, as it identifies property with an object only and further thinks of this object as more or less monolithic, or at least homogenous. This monadic understanding of the word extends beyond real estate or everyday language. For example, when Marx (1991) speaks about the ownership of the means of production, it is easy to imagine them as physical objects that can pass from the hands of the capitalists into those of the workers, without undergoing a transformation themselves. When Jens Beckert (2008) or Thomas Piketty (2014) write about inheritance and wealth, they refer to property – money, stocks, companies – as something that can be added up, measured and passed on to others like a thing.

This understanding of property as a thing is something that many legal scholars from take issue with. They insist that property should not refer to an object that is subject to claims of ownership, but to the rights that legitimize such claims (Hohfeld, 1920; Macpherson, 1978, p 2; Hann, 1998b, p 4). In this view, property is thus not a house or a field, but rather the rights you have been granted in them. Put differently, property is not an object here, but what connects a person to a thing and establishes the former's ownership over the latter. Where common language calls objects property even in the absence of an owner, this view insists that property is a right (or collection of rights), raising the question of where these rights come from and how they relate to the objects we commonly call property. This framework conceives of property as dyadic, not as monadic: it presupposes at least one owning and one owned party.

The disagreement does not end here. Another critical tradition, widespread especially in legal anthropology but also to be found in other fields (Graziadei, 2017, p 73), questions both the monadic and the dyadic model of property. Contrary to the two former schools, this tradition argues that property is not first and foremost – if at all – about things. Chris Hann sums up this stance as '[p]roperty relations can only exist among people' (Hann, 1998b, p 5), citing John Davis: 'you cannot sue an acre; a boundary dispute is not a dispute with a boundary. The study of property rules in general, and of land tenure in particular, is the study of relationships between people' (Davis, 1973, p 157). If we follow this argument, property is not *in a thing* or *in a person* but inherently social: it can only exist *between people*. Land is indeed a very good object to think with in this tradition: generations of legal anthropologists have studied how communities relate to land, how they distribute, negotiate and distribute access and usage among themselves (von Benda-Beckmann, 1979; Vasile, 2007; Strathern, 2009; Dobeson, 2019; Rakopoulos, 2022). The contested nature of land as a property object highlights its social dimension (Dobeson and Kohl, 2023): it is almost impossible to overlook that communities are shaped through and around it. This concept of property-as-relation inherently carries two critical impulses. Like the property-as-right model, it rejects the identification of property with a thing, emphasizing its relational quality. In contrast to the former, however, it localizes this relation between people, rather than between a person and an object. Instead of being one of the poles of the relationship, the object is understood to mediate it (Wiber, 2015).

Each of these three models has its merits, and neither of them is without flaws. Simplistic as it may appear the monadic, 'commonsense' understanding of property captures a material dimension that at times escapes the other two, and it does so with great immediacy. If we want to discuss matters of distribution, inequality or access to property, we will in one way or another have to assume that property is something out there that can be owned. What is at stake in this case is not to deconstruct property but to redistribute it. However, and this is what the other two models criticize, the monadic concept of property both naturalizes things' status and hides the subjective side of ownership. There is an almost naïve assumption that subjects and objects can be clearly distinguished in property, and that the rearrangement of property relations, for example through redistribution, will not have an impact on either of the two.[3]

In contrast, the dyadic model of property-as-right foregrounds the contingency of property and the ways in which people and things are entangled through law. Here the emphasis is on the relation between subjects and objects, and conveys a sense of their asymmetry. Both have a history, which the dyadic model understands as a product of the legal link between them: owner and owned have to be made through law. In doing so, however, it runs the risk of neglecting the materiality of the property object as well as the social nature of the property subject – there is more to property than just rights. In deconstructing property as a thing, property-as-right sacrifices much of its ability to speak about matters of distribution. At the same time, by giving primacy to law it often loses sight of those who are simultaneously in and out of the legal picture.

This is what the third model of property as a social relation can address. Rather than asking how property should best be distributed or how it is legally constituted, it allows to highlight the role it plays in communities, how it divides and connects them, and is in turn divided and connected. To understand property, we must understand its 'social life' (Appadurai, 1986), that is, its social embeddedness and the customs, practices and social structures around it. Such a view goes beyond questions of distribution raised by the property-as-thing model. The critique that property-as-relation enables is not first and foremost concerned with doing away with inequality but perhaps with capturing a particular *culture* of property.[4] Although it does not seek to deconstruct the role of property objects in the same way as the property-as-right model, property-as-relation has a similarly marginalizing effect on things. Far from their primacy in property-as-thing thinking, they recede to the background and become the reference point for dealings between humans, which are understood to be the real essence of property. The critique inherent in this triadic concept of property could be said to be an inherently sociological one: 'it's society, stupid!' While objects' materiality is not denied altogether, a common effect is nonetheless a strangely muted role of things in property, which at times manifests in a 'nothing to see here' attitude.[5]

Arguably, this taxonomy of property concepts is selective and overly simplistic. All good analyses of property are mixtures of these models rather than ideal embodiments of any single one of them. Nevertheless, I would insist that there is truth to them insofar as thinking about their respective ontologies, their strengths and their weaknesses allows us to identify and characterize texts and arguments in the vast literature on property. The vast majority of works on property relies on at least one of these models (and those that do not become all the more interesting for that matter). Although I have presented them here as each turning the weaknesses of their alternatives into strengths, I do not think that they are mutually exclusive. Rather, and this is what I want to explain in this chapter, they can be synthesized in a common concept that allows to grasp the materiality, legal histories and contingencies and the social life of property. The aim is to pay equal attention to the things Rousseau (2011, p 79) lists in his famous dictum on property: the proclamation, the gullible crowd and the fenced-in land.

Revisiting the bundle of rights

Let us start with the proclamation:

> The landowner shall cede agricultural land with an acreage of _____ hectares for the period from 2021/08/01 through 2022/08/31 for conducting the trial described in the following:
> Winter wheat performance trials as well as cultivation of segregating generations in single row plots

The owner of the field shall be responsible for the cultivation of the agricultural land in agreement with the trial manager.

The landowner shall receive the following compensation: 'a. remuneration for the cession of the experimental area and for the related loss of use, as well as basic soil cultivation and basic fertilization, if necessary, €____. b. maintenance measures going beyond this, if necessary, will be remunerated according to common machinery renting rates'.[6]

What is stated, or 'proclaimed' here, is an agreement between a farmer and a public research institute, somewhere in the south of Germany. Every year, the research institute leases a certain acreage of land in the region from farmers to conduct experiments and evaluations with different wheat varieties and breeding lines. After the harvest, these plots are returned to the farmers, who will then use them for their own crops or lease them to others in the following season. This arrangement has advantages for both sides: the farmer receives a secure revenue with minimized input, and the research institute does not have to care about what happens with the land after the trials are over.[7] Land only changes hands partially and temporarily here. There are still things the farmer is expected to contribute to it (such as preparation and maintenance) and the research institute can only use it for the purpose of planting its experimental material. Once both parties have fulfilled the liberties and duties described in the contract, the institute's claim to the acre will cease and it will fully return to the farmer.

Who is the owner here? The straightforward answer is 'the farmer', but perhaps things are more complicated. While in state registers the farmer will still be listed as the owner (Sommer, 2022), at least for the duration of the contract she cannot exploit her land as she pleases. But neither can the research institute claim full ownership: its right to use the land is limited by the terms set in the contract and will end after 31 August. This is not a sale of property, which legal theorists often refer to as 'alienation', although the relationship between the farmer and the institute is certainly transactional and of a business nature. Nor is what we are dealing with here a dispute over or politization of property, but an attempt to distribute, divide and utilize it in mutual understanding. How do we make sense of this messy arrangement?

Legal anthropology and Common Law legal theory have long relied on a particular property concept to analyse such cases. It assumes that in a legal sense, ownership and property are not unified but rather composed of a *bundle of rights and duties* (von Benda-Beckmann, 1995; Turner, 2017; Orsi, 2021). Following William Blackstone, forebear and antagonist of modern property studies, in defining the right of property as 'that sole and despotic dominion, which one man claims and exercises over the external things of the world, in total exclusion of the right of any other individual in the universe'

(Blackstone, 2016, p 42), we all too often take the exception – one person holding all the rights in a thing – for the rule. Instead, we should speak of 'rights' in the plural, since property theory recognizes a range of different rights that different owners can hold and exercise, such as the right to use, the right to alienate, the right to exclude and so on (Schlager and Ostrom, 1992). We also should not overlook that in its early days, theorists took care to add 'duties' to the 'rights': unlike liberal and libertarian philosophies of property, which gained influence in the second half of the 20th century (Radin, 1993), the *bundle* concept recognized that property came with both freedoms and obligations (Maine, 1908, p 158).[8] Wesley Hohfeld (1920) considered rights and obligations to be each other's flip-side: where rights introduce degrees of freedom, allow for movement or grant privileges, duties attach ties and obligations, restrict an owner's use of an object and demand back compensation (Tellmann, 2022). Rights push and move where obligations pull and fix.

The *bundle* is not without criticism. The most widespread accusation against it is that it runs counter to the commonsense, monadic idea of property as a thing, dissolving the latter in a myriad of rights and duties (Grey, 1980; Penner, 1996). The result, its critics rightly point out, is that the category of property increasingly becomes blurry and indistinguishable from other legal institutions – for example, marriage, tutelage, public office – that are no less collections of rights and duties (Hann, 2005). There is also the problem that what might legally be dissolved in a multiplicity of sticks remains physically very much unified: we are still dealing with a continuous area of land, no matter how strongly we emphasize its legal deconstruction.[9] But from a pragmatist point of view, the advantage offered by the *bundle* is its ability to move back and forth from 'full' or 'Blackstonian' property to arrangements in which we are dealing with multiple, partial and distributed owners and property objects. Edella Schlager's and Elinor Ostrom's (1992) famous study on property-rights regimes remains an instructive example for how to do this: by starting from full ownership and subsequently removing stick after stick to arrive at more nuanced property relationships that different members of a community have with a shared resource such as a fishing ground.[10]

If we apply the bundle of rights and duties to this contract, we can read the agreement as a distribution of legal sticks between the farmer and the research institute. The text attaches a number of different rights and duties to the piece of land on the one end and to the two owners on the others. The landowner is tied to the land via the right to compensation and the obligation of basic management and fertilization; in turn the research institute is granted permission to conduct trials and given the duty to pay the farmer. Beyond these provisions, the farmer retains the remaining freedoms and obligations of the property bundle; the temporary transfer of the use right in his land will also return to her after the lease.

This is a very simple property arrangement. We could easily think of much more byzantine cases in which agreements are struck with multiple parties, who might even transfer their sticks further.[11] A typical example is an apartment complex in which the apartments are privately owned and let to renters, while the house belongs to the community of owners and is managed by a facility management company acting on their behalf. The company will take over many of the rights and duties tied to the complex from the owners and in turn contract these out to cleaning services, construction firms, maintenance workers and so on. Meanwhile, where the original lease contracts permit (or do not explicitly prohibit), renters might sublet 'their' apartments or take in housemates to share their apartment with others, a form of cohabitation often governed by cleaning schedules and neatly partitioned fridge shelves, and so on.

The conceptual advantage of the bundle of rights and duties is its ability to keep track of these entanglements and map them without having to clearly demarcate one owner and one property. Property in the bundle of rights and duties is divisible, which means that we can use the concept to deconstruct Blackstone's absolute property. At the same time, it is also modular, which allows us to do the opposite; combining rights and duties until we arrive at a comprehensive bundle that bears striking resemblance to full ownership. It is easy to see why the *property-as-relation* school especially is so fond of the bundle as a concept: it is well suited to the messy reality of real-life property arrangements. But is there also something that, despite its sensitivity to legal complexity, it remains blind to or ignorant of?

From rules to practices

If we think of property as produced through and governed by legal texts, one of the first things we will notice is that these texts are rather thin and incomplete. Not everything can be spelt out on the pages of a contract. Consequently, things like the exact way to manage the field are left to 'accordance' and 'consultation' in our contract. Similarly, a cleaning schedule in a shared household usually determines when someone will be in charge of cleaning but does not define how the cleaning will have to be done. We could think of laws and contracts as very rough screenplays or storyboards (Latour, 2013, pp 389–390): they can give some directions for acting but not replace or micro-manage it. What is more is that a large share of agreements over property are never written down but remain verbal only, are never made explicit, or perhaps not even conscious in the first place. In the case of the land lease, for example, the contract is supplemented by a handshake agreement over the harvest: the institute will only harvest and retain the seed it deems worthy of further propagation and multiplication. This leaves the rest of the wheat on the field to the farmer, who can keep the underperforming part of the harvest and sell it as animal feed.

This part of the agreement would escape us if we only looked at formal legal texts as our data. To account for the role of 'informal' property ties, we also have to speak with the people involved and inquire how they go about honouring the law in practice. The *bundle* tradition has not been oblivious to this fact: anthropologists, economists and political scientists have long sought to highlight the informal side of property arrangements to paint a fuller picture (von Benda-Beckmann, 1995). One way of accounting for the informal side of property is to speak of 'rules' instead of 'law', an approach favoured for example by Ostrom, Gardner and Walker (1994, p 38). Rules refer to everything that governs our conduct with things, including what happens if we break them. Such a concept resonates more easily with the verbal, unwritten and even the unconscious, as Pierre Bourdieu (2008) has shown in his ethnographic studies in Kabylia: people are often unaware of the rules that attach strings to things or make them behave in certain ways but not in others. Nevertheless, an outside observer can still discern a set of rules and norms these people adhere to and distil a coherent logic from it. Obligations become visible, for example, when people feel the urge to give back something to someone else, such as returning a property item to its rightful owner.

And yet Bourdieu's perspective also shows a problem with rules. If we cannot predict how people go about property from explicit rules, but instead have to infer these rules from people's behaviour, what exactly is the relationship between rules and practices? For Bourdieu, practice is a rather inert, structured and collective activity: the people who partake in it can be exchanged without the system of practice collapsing. It does not matter for a property regime in housing, for example, who exactly lives in a given apartment; nor is the landlord's name of particular importance – the rules for renting and letting will be the same. This approach can only connect practice to rules if the former is a recurring routine, something for which there is already a script, rather than something that has to be figured out first. But how exactly do I fertilize a particular field? How exactly do you plant a row of experimental plots with a selection of wheat varieties? To assume that for all of this is there are already rules 'in practice' would overlook the work that goes into fulfilling a contract, as well as the ambiguity and uncertainty that comes with it.[12]

An alternative could be to think of rules and practice as complementary, rather than synonymous: here a more general rule would set the frame for something that has to be specified in practice. On one side we could have a *bundle of rules* and on the other there would be a corresponding *bundle of practices* (Braun, 2021a). The rather general right to plant and harvest experimental plots stated in the contract, for example, would translate into a very specific and elaborate activity on the field, involving precision and planning that have no place in a legal text. In more general terms, we can add at least one practice stick for every rule stick: for the right to exclude, there is a practice of exclusion; for the right to alienate, there is a process of alienation; for the duty to compensate, there is an act of compensation. But between them we can also add those practices that have no equivalent in the law, those that go unmentioned or are presupposed without being specified in the text of a contract. Just like the bundle of rights and duties, the bundle of practices can be divided, reassembled and distributed between several parties: if anything, adding practice to the picture shows how difficult it is for someone to hold the full bundle in their hands.

To speak of a bundle of practices shifts the focus from *having* to *doing*, from the privilege to the work of property, as John Page (2017, pp 201–210), Nicholas Blomley (2004, pp 1–28; 2013) or Johannes Schubert, Stefan Böschen and Bernhard Gill (2011) have argued. One way or another, an emphasis on practice carves out a space of investigation that social sciences can consider 'their' playing field (Schatzki, Knorr-Cetina and von Savigny, 2001; Soler et al, 2014), creating a counterweight to the realm of laws and legal texts claimed by legal studies. Rather than deconstructing or contextualizing the legal side of property, a focus on practice immediately shows what disciplines beyond law can contribute to the study of property. Property emerges as performative: it does not come into the world simply by proclamation (Rose, 1985) but through practice (Blomley, 2013). Broadening our focus beyond rights, duties and rules is good because it draws our attention to the difficulties that property encounters in practice and makes room for contingency and nuance: the gaps left by the legal text of the contract can be filled and modified by social practices.

A technical matter: towards a bundle of scripts

For the same reason, however, the juxtaposition of rights and practices is also a problematic one. Either we risk overstretching the concept of rules, overemphasizing the static, repetitive aspect of practice, or we create a theoretical division between rules and practices where empirically we often find a continuity. Both the original bundle of rights and the bundle of practices, which either extends or complements the former, subscribe to a rather one-sided idea of property: they conceive of property as something to be enacted, as a flow of agency that starts with the drafting of legal texts, which are subsequently translated into and refined by social practices. The problem with such an idea of practice and performativity, as Nicholas Blomley (2013) rightly points out, is that it still a little too intentional, still a little too social constructivist,[13] a little too anthropocentric. But what about the work of fences?, Blomley asks. Can we understand their contribution to the enactment, stabilization and reproduction of property as intentional? Does it make sense to call what a fence does 'practice'?

Or, alternatively, 'rules', 'rights' or 'obligations'? Where does a fence fit in our two bundles?

Blomley's answer is to include the fence in his concept of performativity, following the work of Michel Callon (1998a), Donald Mackenzie (2006) and others (Mackenzie et al, 2007; Muniesa, Millo and Callon, 2007) who have explored the relationship between economic knowledge and market devices. 'A fence performs property when it is hooked up to other entities. Put more generally, property is performed when such entities stabilize and work together', Blomley (2013, p 39) formulates it. Property, he adds, is an 'assemblage' of very different elements: laws, practices, fences, registers, and so on, each making a contribution to the whole. Following Blomley, we could thus add a third type of components to property next to rights and practices, which we could call *technologies* or *devices*. These come in where legal and practical elements fall short: laws against theft may be toothless, guarding your possessions 24/7 may be impractical, but a bike lock will firmly and patiently hold on to your property.[14]

The addition of yet another bundle – that of technologies – to those of rights and practices thus seems necessary. Like the two other bundles, it would allow us to classify technologies with respect to what they contribute to the whole of property. Fences and door locks could be understood as technologies of exclusion, fish nets as means of appropriation and 'no trespassing' and 'private property' signs as devices of proclamation. At the same time, dividing property into three separate bundles seems less helpful if we agree with Blomley that laws, practices and technologies have to work hand in hand, rather than separately, to weave and hold owners and possessions together.

How else could we then articulate property without overlooking any of its diverse constituents? Is there a way to rearrange the three bundles in a way that mixes technical devices, social practices and codified rules? In a slightly dated piece, Bruno Latour offers an answer, ironically discussing a classic property problem without mentioning property a single time (Latour, 1990).[15] A hypothetical hotel manager is in trouble over retrieving his key from his guests. How can he make sure his property will return to him? So far, his efforts have yielded little success: all too often, his customers forget to return the room keys when they leave the hotel. The result is that they lose them on a stroll through town or end up taking the key back home with them, which causes a lot of trouble for everyone involved. The explicit request at check-in to return the keys helps, if only a little, and even the addition of a sign reminding them to leave the keys at the reception cannot make all off them circle back to the hotel manager. It is only the addition of bulky, heavy weights attached to the keys via metal rings that suddenly makes guests return their keys: almost magically, they seem to have lost any desire to carry them beyond the premises of the hotel.

31

Latour's example may be fictional but is nevertheless instructive.[16] What counts in the end is less what the elements in the hotel manager's strategies are made of – written letters, spoken words or cast iron – but whether they make the keys return to him or not. Eventually, signs, pleas and metal weights all work together to turn the hotel keys into well-functioning property that returns to its owner even if it has been given away temporarily.[17] The argument Latour makes here is not simply that property is heterogenous and consists of many different elements beyond an owner and her object. What matters, he stresses, is not the nature of these elements but the specific effects they produce together. Likening them to computer programs, he calls these effects, which are relayed across chains of practices, rules and devices, 'scripts' (Latour, 1990, p 111). The term, like the related concept of inscription, has a long tradition in the sociology and anthropology of technology (Knorr and Knorr 1978; Johnson, 1988; Akrich, 1992; Latour, 2013, pp 390–409). Following loosely in the semiotic tradition of A.J. Greimas (1983), both have been used to talk about agency that does not originate in but rather flows through objects and subjects: the hotel guests and keys do not act so much as they are *made* to act in a particular way.[18]

Inscribing property into things, from lunch boxes to cattle skins, is a common practice. It serves as an enduring reminder of ownership, as testimony of bestowal, as a perpetual reenactment of Rousseau's proclamation: 'Lucy's', 'To P, with love', 'Private Property, No Trespassing!' But inscribing property is not merely a more iterative form of performance in Blomley's sense. In Latour's example, a particular feature – weight, and thus inertia – is inscribed into the hotel keys through a metal ring and a heavy, bulky attachment. The result is that the keys start to behave more like well-trained dogs than like stray cats; diligently returning to their owner. Under the concept of property as a bundle of rights, we would have to take this outcome for granted: after all, the keys rightfully belong to the hotel, just like dogs belong to their owners. Under the bundle of practices, we would instead have to explain the successful recall of property as the guests' doing. In contrast, a bundle of technologies would only look at the key rings and the weights.

The key message of Latour's example, however, is that we should not understand agency and causality as originating in one single point or being confined to only one type of entities – rules, persons or objects. What makes keys return to the hotel is a script that runs through the owner, rules, pleas, metal rings, weights, narrow pockets and the guests' sense of convenience (Latour 1990, p 110). The script could be articulated differently: through more diligent and attentive guests, through a porter who checks every outgoing guest for keys or by the increasing use of reprogrammable key cards that can be issued and 'withdrawn' with a few clicks at the reception computer. Especially this recent re-articulation of the problem illustrates how scripts run through artefacts, rather than being strictly tied to them. A bundle of scripts thus ties together a long list of very different elements. One of the positive side-effects of scripts, however, is that they resonate with each of the three components discussed in more detail in this chapter: in their sense of texts, they include the letters of laws and contracts; as instructions for actors and narrative structures of a play, they shine a light on practices and performance; in their meaning as self-executing pieces of software, we can understand them as technical elements endowed with autonomous agency.

Scripts in action

Like rules and practices, scripts are diverse. Return upon recall, or re-appropriation, is just one among many key issues that put property to the test. As we will see in the following chapter, the alienation of property requires very different tools, practices and legal foundations; the successful exclusion of others from an estate relies on yet another set. Some scripts unfold within the confines of a hotel, others extend beyond jurisdictions or span several decades. At the same time, they might rely on the same laws, the same practices, the same technologies. Moreover, property is of course not the only thing we could understand along the lines of scripts: organizations, theories, transport systems or first dates could easily be described as consisting of scripts, too.[19] So how can we tell different scripts, different forms of property and other social forms apart? One last time, I want to return to the plot lease here to give an answer.

Early August. Harvest time. The whole wheat research department, including interns, summer jobbers and myself, is on alert. It has been raining for the last few weeks, leaving no window of sun to bring in the harvest. Today, however, all of the staff are set to harvest the last remaining field, a few miles from the research institute. Over the past months and weeks, they have evaluated every single row and plot, marking each of them on a tabular scheme of the field, and they have examined the data to decide which ones warrant harvesting and which ones would be discarded. In accordance with the handshake agreement, the institute will only keep the plant material that it deems worthy of further propagation, while the farmer gets to take the remainder to the mill or the silo: render unto the farmer the things that are the farmer's.

The plots with the promising breeding lines are not neatly separated, however. On the field, they are intermixed with those of the underperformers. So how are the terms of the agreement to be satisfied? The plan the department has come up with involves two harvesting machines of different sizes, a truck and a small army of farmhands. The smaller one of the harvesting machines, matching a single plot in width, will thrust from the road into the field. It will thresh those plots in the first row from the field that have been marked for harvesting on the plan. The farmhands will quickly take the bags filled with grain from the machine, tie them up, put a unique label on them and dump them by the roadside. In a pincer move, the big machine will then advance from the side of field and, in perpendicular direction, take care of the bad plots. Once the first rank has fallen, the tactic will repeat with the second, and so on.

And indeed, this is how it happens. The smaller blue machine charges into the first row, pulls back onto the road, takes a turn to the right, drives forwards until it reaches the next prospective plot, turns left and thrusts into the row again. Meanwhile, the farmhands are swarming around it, receiving small white bags filled to the brim with grains, tying them up with one swift move, putting the plot label on the cord, dropping the sacks by the roadside. Once the first row of plots has been interspersed with breaches, they all move out of the way. Only seconds after, the big red harvester takes what is left (Figure 2.1).

After the first row is cleared, blue harvester and farmhands turn to the next one, selectively taking out plot for plot, until the rest of the row is mowed down again. The sacks are put on the trailer and taken to the institute to be stored in the drying chamber, while what the big machine has harvested is going the way of all grain. The whole operation is a complex choreography of men, wheat and machines, impossible for me to capture with my poor phone camera. But by the end of the day, the good are in the pot and the bad in the crop; what used to be an unsortable mess of roots, stems, leaves and grains has been divided up between the farmer and the research institute: the contract has been fulfilled.

What the institute has come up with here is a script for distributing an object between two prospective owners. The handshake agreement divides the wheat in one part that belongs to the institute and one part that belongs to the farmer. Both have very different destinations: the institute's share

Figure 2.1: Two harvesters splitting up property

is to be sown again for the next year, the farmer's share will most likely be turned into fodder grain. What the agreement does not answer, however, is how the disentangling of the two wheats is going to be achieved; nor does it actually bring together owners and wheats. This is because the agreement, when it is struck, does not know what the future is going to look like but also because its words are different from the wheat in the field. All it does is provide the institute's staff with a very loose script for producing two separate property objects (Latour, 2013, pp 390f).

This loose script is not enough, however, for getting the job done. A more refined script is needed, the contract's terms have to be elaborated in practice. The first challenge is dividing up the wheat. For this purpose, the breeders at the department go out in the field, mark the plots they deem worth keeping and transfer these marks into a map of the plots. Even though the map is strongly abstracted from the field, it retains essential characteristics of its geography, such as the gaps between individual plots (which serve as boundary markers) and the relative arrangement of the plots (such as sequence and rows). This means that the abstraction is only possible because the plots have been sown in a particular fashion that allows the translation from field to map to retain what is important and to leave behind what can or should not be translated (Latour, 1987; 2005, p 106f).

But it is also what allows the map to turn into a script as it can be re-translated to the field: the sequence and relative positions of the plots on the map are the same as in the field. Now everything that is needed is a technical apparatus able to follow the script and pick up the two different kinds of wheat. This is a slightly different matter than division, it is primarily one of *appropriation*. The technical apparatus consists of a harvester small enough to harvest just one single plot of wheat while leaving the ones next to it intact. The harvester must not be much wider than a single plot but not narrower either, otherwise it would seize either too much or too little. It separates grain from straw, channelling the former through a pipe into bags held by the farmhands. This opening is a crucial element because the harvester otherwise could not *pass on* the wheat to the farmhands. The bags are part of this sub-script that passes on seed from the harvester to the trailer, but inside their fabric, they also *contain* the wheat, holding it together so that it does not get lost on the way.

Only after the small blue harvester has appropriated what is of value to the institute the bigger and less precise harvester can take what is of value only to the farmer. This is again a matter of division, solved here through combining the smaller harvester's precision with chronological sequence. The bigger harvester's tank allows it to contain the grain without having to pass it on in between; at the end of the day it will fill a whole trailer with its load. Both trailers will head to different destinations, finally *rejoining* the owners with their property as designated in the contract.

We encounter a series of operations here that are connected to each other and yet follow different trajectories. The first movement is bracketing, as Blomley (2014) calls it, or *division*, which delineates and splits up the object. It is performed to a succession of steps that inscribes a difference into the wheat field. The next is a matter of *appropriation*, which seeks to gain hold of the wheat. We can understand the mapping and translation of plot data as acts of appropriation already, but physical appropriation only begins with the movement of the harvester into the wheat. Appropriation is not enough, however. The script must also be able to switch from *appropriating* to *passing on* to allow other actors closer to the final owner to appropriate the wheat in turn. Appropriation is always threatened by unintentional transfer: the harvest may get lost along the way. This is why between the intended points of passage it has to be *contained* inside seed bags and storage tanks. Finally, the wheat has to be rejoined with its owner by returning to them what is already theirs on paper and which they have only let go of for the term of a harvest contract.

How do we know that all these processes are part of property? And how can we tell that others, like putting labels on the seed bags, which identify the harvest of a certain plot, are not? Property scripts revolve around producing property and owners and establish a link between them. The wheat in the field is subjected to numerous transformations which, step by step, turn it into two different appropriated objects. But not all transformations and modifications serve that particular purpose equally. The labelling of bags does not create a link between the department and the wheat, but between the seed bag, the plot, and the data drawn from that plot. For the aim of dividing and distributing the harvest between two parties, this link is of secondary importance. What characterizes most property scripts is that they make a difference for the proximity or distance of property objects and subjects. Depending on how widely one interprets proximity, we could indeed add the plot labels to the list of property scripts. Labelling establishes an intimate access to the seed bags, one that will allow the department to treat them in a much more controlled and informed fashion than bags without labels. These labels connect the values in the records to the seed in the bags; if they go missing, the bags will lose any value for the department and simply be turned into fodder grain as well.

The question of what contributes to property is thus a pragmatic one (Radin, 1993, pp 1–34): it is contingent on the issue and the relations taken into account. For the issue addressed by the harvest contract and the triadic relations between farmer, department and harvest, the bag labels do not make a difference. For the dyadic relationship between the seed bags and the departments, however, they matter a lot. Speaking about property does thus not just require a conceptualization in which all parameters are defined from the outset. If we want to approach property pragmatically, we have to first and foremost ask for the issue, the *pragma* that actors and actions seek to deal with by means of property (James, 1898).

The matter at stake in this example is a wheat harvest and its distribution. While there are no disagreements about the terms of its distribution – the good in the bag, the bad in the tank – the logistics of the agreement prove challenging. Looking closer at how the contract is performed and its scripts extended, we find new issues that stand between the contract and its satisfaction: property objects have to be demarcated within the undifferentiated wheat, two demarcated property objects might mix if the harvesting party does not pay due attention, property might be lost or stuck along the way if containment and passing on do not harmonize. At each of these crucial points, we find sub-scripts intersecting with and running parallel to the trajectory of the division script. Whether following their trajectories contributes to our understanding of property depends on the issue we intend to address.

Conclusion

From the initial question of this chapter – how to conceptualize property – we have come a long way. My aim in this chapter was to broaden the idea of property to make space for disciplines other than legal studies and to offer common ground for contradictory models of property. Divergent as they might seem, I believe there are ways to reconcile and equally use them as conceptual resources. The popular idea that property is a thing, problematic as it may be from the point of view of critical property studies, has its rightful place in property theory (Grey, 1980): it reflects our encounters with and expectations of objects that behave like property. Such objects, like the keys in Latour's hotel, indeed exist. But things are not born as property, as critical studies of property have pointed out time and again, they are *made* property. If we want to understand why they obey and return to their owners, we have to take them apart and put them back together again.

Property-as-right highlights this constructed nature of both owners and owned. Property, this model argues, has to be instituted (Pottage, 1998). Indeed, legal provisions, agreements and fictions (Fuller, 1970) play a crucial role in delineating the elements of property and providing them with particular agency: an active, responsible, sovereign agency for the owner; a passive, reliable and docile agency for her possessions. But in themselves, they are not sufficient to make property, leaving a lot to imagination and even more to practice. Beneath the legal tip of the iceberg, there lies a whole world of practice (Navaro-Yashin, 2009; Beck, 2016); one we all too often overlook when we pick laws and rules as the privileged point of departure for a critical analysis of property. We can imagine doing property in the absence of property law (Serres, 2007) but we will struggle to find property law that is not instituted and enacted through practice.

If I nevertheless do not speak of property as a bundle of practices throughout this book, it is for two reasons. First of all, I believe the divide between the legal and the practical is a misleading one: there are practices in the law (Pottage, 1998; Scheffer, 2010; Umphrey, 2011) just like there are rules, explicit and implicit, to practice (Bourdieu, 2008). Second, a dichotomy between rules and practices would suggest that their combination would give us a full picture of property. But as we have seen, there are elements of property that cannot convincingly be categorized one way or another. The example I have used here is technology, an element that features prominently in the following chapters and has caught the attention of others before me (van Dooren, 2007a, pp 137–170; Perzanowski and Schultz 2018, pp 139–154). Unless we want to add a third bundle to rights and practices, turning the dichotomy into a trichotomy, I believe we are better off leaving open what exactly property is made of in each instance: this is a matter for inquiry (Latour, 2005).

Speaking of property as a bundle of scripts is meant to do exactly that. It is intended to redirect our gaze from the particular embodiment of property to its aims and ends. Scripts can unfold in very different ways even when they pursue the same goals or perform similar functions. In studying the articulation of property, we should thus not just look at what it is made of – we should also ask which scripts each of its elements serves and follow them to learn where they reconnect or go astray. The image of the bundle is convenient, as it will let us understand some of the dizzying entanglements of property that we find in the seed sector without resorting to the overly simple juxtaposition of private and common property. Yet the bundle also reminds us that it is not guaranteed that the various scripts of property will eventually work together, rather than diverge or contradict each other. As we will see in the next chapter, fine-tuning a bundle of scripts is cumbersome undertaking and only rarely crowned with success. The notion of scripts in turn suggests that there is an agency to property that exceeds both the active role of owners and the passive role of property objects. If the crises of property are multiplying, it is to no small part due to the agency of property scripts that, once instituted, cannot easily be accommodated in the same bundle.

The challenge for the remainder of this book is thus to find out scripts work, fail to work, or do more work than expected. What allowed a particular script to proceed in the past? And why has it ceased to function? How come plants once lent themselves to being propertized whereas they refuse to do so today? And can the problem be fixed? These are the questions I will try to answer in the following chapters.

3

Property and the Market

We stand. And wait. The door is shut: inside, the examiners are discussing. We, outside, are biding our time with jokes and pointless second-guessing. The casual tone cannot diffuse the tense atmosphere in the otherwise empty hallway. 'What were they aiming for with that comparison?' Hubert wonders. Wolfhard thinks that in the future, they should use least-square means data[1] only as last resort. Sabine complains that the bar is set too high this year. And I listen and scribble. On a day in March, we have travelled to Hanover, some 400 miles from Hubert's breeding station, to attend the annual variety hearings at the Federal Variety Office (*Bundessortenamt*, BSA). We are joined by Hubert's colleague from Lemgo, Sabine, and his head of marketing, Wolfhard, who have come along to answer to the office's examining board. Sabine, Hubert and Wolfhard have cautiously prepared presentation slides meant to demonstrate the superiority of each of the five varieties of wheat to be examined today. In the half hour or so between arriving at the office and being called to the hearing room, they have nervously rehearsed their argumentative strategy and run through its strong and weak points. While we were sitting through the hearings, they have fiercely advocated for their varieties and tried to rebut all of the examining board's criticisms. And now we stand. And wait.

The verdict we are waiting for is one that will determine the future of Sabine's, Hubert's and Wolfhard's wheat. What the Federal Variety Office will decide that day is whether it will grant them an at first sight minor although crucial stick for their property bundle: the right to sell their seeds. In strict terms, this is not even a property right. German law distinguishes between intellectual property in seeds, so-called PVP (*Sortenschutz*), and market admission of varieties (*Sortenzulassung*). And yet without the latter, the former is worthless. If the Federal Variety Office rules that the varieties under examination are fit for the market, Wolfhard, Hubert and Sabine can go ahead and sell them to growers all over the country – if the examination board arrives at the opposite conclusion, the seed already ramped up in storage will be turned into animal fodder instead.

This is certainly a strange arrangement, especially as the Federal Variety Office is also in charge of granting breeders property titles for plant varieties. In this case, each of the varieties has received variety protection, meaning that Sabine and Hubert have been already given their intellectual property. Why, then, does the same institution go through another bureaucratic process if it has handed out the bulk of the property bundle beforehand? Why is there a distinction between property in varieties and admission of varieties? And why do the three think they need plant variety protection in the first place?

To answer these questions, we have to take a look at plant breeding as a professional industry – an industry of central importance for the agricultural sector and beyond, but at the same time a surprisingly small, young and low-tech industry. Its key product and source of revenue is seed, which comes in countless forms and sizes: from minute lettuce seeds, to seed potatoes and other tubers, to cuttings for fruit trees and wine. This seed is produced for large-scale agriculture and greenhouse horticulture, for sowing grass strips on roads as well as growing ornamental flowers. Although plant breeding is often portrayed as one of the oldest practices in human history, reaching back to the very beginnings of agriculture, its modern form can be dated back roughly 150 to 200 years (Acquaah, 2012, p 22). Breeders will often distinguish 'old' from 'new' plant breeding by the latter's discovery of cross-fertilization and genetic knowledge in the 19th century and its subsequent reliance on them (Becker, 2019, pp 12f). At this point in time, the popular narrative goes, breeding and farming became two separate, rather than synonymous, activities (Acquaah, 2012, p 23; Bonneuil and Thomas, 2012, p 17).[2] In this chapter, I am offering an alternative distinction: modern plant seed is seed that is sold on markets, meaning that it can be alienated and appropriated in particular ways; with profound implications for plant breeding as a practice.

Due to its nature, seed poses a challenge to propertization. Under certain conditions it behaves like a tangible, exhaustible mass commodity, not much different from nails, shoes or automobiles. In other circumstances, however, it becomes more similar to intangibles like music, literary works, or software; a potentially endless resource that can be multiplied and shared (Gill et al, 2012, p 420). How do plant breeders and farmers resolve this contradiction? The answer, as will see, is that they do not: seed is a servant of two masters. Although it may at first appear like an ordinary market good, it remains what Nicholas Thomas (1991) calls an 'entangled object', tied to two owners even after it has been sold. What is remarkable about this arrangement is not that is unusual. Quite to the contrary – it sheds light on the property condition of market objects in general.

Forming seed

Hubert Kempf is a wheat breeder of the old school, although he would likely protest against such a characterization. From spring to fall, he sacrifices his weekends to travel across the country and spend hours in scorching sun and pouring rain, inspecting plot after plot, row after row, plant after plant, and filling his notebooks with numbers.[3] Hubert values hard work and prizes quality over quantity. At times he despairs over younger colleagues' love for free time and calling it a day early. At others he shakes his head over the exclusive obsession some of his competitors and customers show with yields, rather than his appreciation of a well-rounded, resilient variety. And yet, with his tanned face and stern façade, he is more of an archetype than an antithesis to the average breeder. If there is something that all the various branches of plant breeding have in common, it is that they rely on intensive and extensive labour, especially during the summer months.

Hubert leads the regional subsidiary of a medium-sized seed company near Munich specializing in wheat and barley. He has been working as a wheat breeder for over 30 years and overseen the sale of the wheat program from a family-owned firm to a French barley breeding company, Secobra Recherches, as well as a rapid expansion of its market share in the last few years. Secobra is not the only French company to have entered the market; its competitors RAGT and Limagrain, too, have successfully gained a foothold in German wheat breeding. Next to them, there are also established domestic players like KWS Saat, DSV, Nordsaat or Strube, as well as more recent ventures by biotech and agrochemical companies like Bayer, BASF or Syngenta. Germany is the second-biggest seed market in the EU after France, the EU single market in turn being the third biggest on a global scale in terms of turnover (Ragonnaud, 2013, p 9). In spite of a continuing trend toward consolidation and market concentration (Mammana, 2014; Bonny, 2017), the German market for wheat seed is still remarkably diverse and competitive. At the time of my research, there are about 17 active programmes, Hubert estimates (Kempf, 2016).[4]

'Modern plant breeding is an art and a science', George Acquaah (2012, p 33) proudly states in his introduction to the field. The truth, at least as far as commercial plant breeding is concerned, is that it is much closer to a craft than to either of the two. This is not to denigrate the work of plant breeders and the achievements of modern plant breeding – quite the opposite: since they cannot rely on calculation or creativity, Hubert and his colleagues have to make up for it with all the more labour, experience and intuition. The development of a commercial plant variety takes about ten years in wheat and up to three decades in fruit trees (Becker, 2019, pp 194–205). The results of plant breeding are largely unpredictable – the vast majority of plants produced in a breeders' nursery will be discarded sooner or later. Despite the frequent emphasis on genetics as guiding science for plant breeding, it has remained of little help to applied breeding (Wieland, 2004; 2006). Due to the random nature of genetic split and recombination, no one can predict the exact outcome of a cross between two plants. To make sure that they are not missing out on a promising combination of genomes, Hubert and his colleagues will perform the same cross-fertilization several hundred times over. This means that the minute flowers of the wheat plants involved in the cross will have to be carefully castrated with tweezers and skilled, steady hands, after which the ear of a pollen donor (the 'father') is bound to that of the recipient plant (the 'mother') – a process repeated several thousand times every year in spring (Becker, 2019, pp 196f).[5]

Conventional wheat breeding can be compared to a blind shuffling of two decks of cards with the aim to end up with a better overall deck: there is no way to reliably say whether you are going to obtain a superior combination if you reduce the shuffled deck to half its size.[6] Hubert's aim is twofold: one is to identify, stabilize and improve individual characteristics in his nursery to later include them in as many plants as possible. In wheat, this so-called trait development is especially important for resistances against fungi and insects. In other plant species, it is of much more central importance: producing orchids of a new shape or colour, breeding lettuce that will last longer on supermarket shelves, finding onions that will not make you break down in tears when you chop them. Vegetables as we today know them would be inconceivable without deliberate trait development: they would wilt in the fields, rot during shipping or perish after a day in the fridge. Trait breeding is easiest where it relies on only one or a handful of genes that will likely be passed on together. It can be likened to searching for particularly good cards and finding ways to make sure they end up in all your decks.

The other goal of plant breeding is to further improve existing genetic decks by shuffling them many times over and hoping that as many different good traits as possible will end up together in the same plant. Deliberately transferring a specific trait is possible in theory but expensive in practice. Although genetic engineering promises to do exactly that, the established procedure is just as random as cross-fertilization – no one can predict whether a gene is going to end up in the right place on the right chromosome (Prado et al, 2014). What is more, genetic modification with gene guns or *Agrobacterium* vectors (Charles, 2001, pp 74–91; Lurquin, 2001, pp 56–102) is vastly more expensive than conventional plant breeding: obtaining a non-transgenic variety is estimated to cost about US$1–2 million (Goodman, 2002), whereas a single successful transfer of a transgene will usually cost in US$100 million and take ten years for research and development only (Phillips McDougall, 2011, p 7; Prado et al, 2014, p 772).[7] Marker-assisted breeding is cheaper and allows breeders to keep track of genes that might not be visible in the field because they are recessive alleles or because they code for resistances against absent pests (Reece and Haribabu, 2007). Its use requires a laboratory and knowledge about gene functions and markers, which only makes sense for a few select and valuable plant traits.[8]

For reshuffling the genetic cards of his wheat, Hubert can thus make use of a limited technological and epistemic toolkit; a toolkit remarkably similar to that of breeders a century ago. One strategy is increasing the number of tries and decks involved, which will lead to a more genetically diverse nursery but at the same time require more work for sifting through the crosses. A single ear of wheat will yield between 12 and 20 seeds; the overall number of plants will thus increase very quickly over subsequent generations.[9] After the initial cross, the wheat plants will self-fertilize, producing ever-more inbred offspring each generation. In the first few years, breeders have to rely on their trained eye to examine the performance of breeding lines, as the number of individuals does not allow for statistical analysis (Timmermann, 2009). The further the breeding process advances, the bigger populations become, allowing Hubert and his colleagues to draw from bigger sample sizes and to evaluate full plots rather than single plants. At the same time, they have to increase selection pressure and weed out underperformers with less mercy if they do not want to drown in plants and work.[10] After all, labour and nursery space are finite resources, and over the development cycle of a variety, the focus will thus shift from increasing genetic diversity to reigning it in.

Although Hubert spends a fair amount of the year out in the fields, the fate of breeding lines is decided on in the offices of the breeding station (usually in the winter months), where the numeric evaluations from his notebooks are transferred to coloured spreadsheets that allow for quick comparisons between the various wheats in the nursery. The more and the darker the green (indicating desirable values) in a row, the better; the redder a particular line's numbers, the closer it moves to being dropped from the programme. Hubert and his colleagues evaluate a number of different properties: the one most sought-after by retailers and farmers is certainly yield, although Hubert insists that pathogen resistance is key to a good variety; a philosophy that often requires painful compromises when selecting. Protein quality, tendency of plants to lodge or resistance to late frosts are other properties that are of importance to farmers and millers (and thus to breeders).

Over subsequent generations, the experimental wheat lines will become uniform, in appearance and genetically. A single plot planted with a pure wheat line at this stage will look remarkably homogeneous and at the same time distinct from the ones next to it (Figure 3.1). In the later stages of the breeding cycle, this is what allows breeders to compare them in the field, to use statistics and to control for environmental variables like weather and pathogen pressure when compiling their data. These data are used for predicting the performance of the various pure lines in the official nationwide trials (*Wertprüfung*) coordinated by the Federal Variety Office. These trials form the basis for the office's decision to admit a variety to the market or to reject it. The purification and homogenization of seed

Figure 3.1: Standardized plants in a nursery

Note: Each plot is occupied by one distinct lineage or variety.

is not just an epistemic prerequisite, however. The final variety of wheat, once it has successfully passed the various cycles of evaluation, will be a recognizable, standardized and thus calculable plant that can be multiplied without much effort. After admission has been granted, seed will first be bulked-up, then shipped, and finally be sold to farmers through agri-trade retailers or breeders' own channels. This is a great advantage for breeders, seed distributors and farmers alike. From the point of property, however, it is a considerable problem.

All too commodified

Seed markets are a comparatively recent phenomenon. For most of agricultural history, plant breeding and farming were synonymous: when harvesting, farmers would put aside some of the fruit or grain to plant it the following year, more or less consciously selecting for the best-adapted plants (Kloppenburg, 2004, p 2; Acquaah 2012, p 23). It was only toward the late 19th century that plant breeding emerged as a distinct profession and activity in Europe and North America, dedicated to the production of seed for others (Bonneuil and Thomas, 2009, pp 29–62; 2012, pp 17–34; Sanderson, 2017, p 22). Breeders began to import foreign varieties, deliberately crossed existing races, selected the progeny for performance and purified it through self-fertilization (Bonneuil, 2006; Wieland, 2006), at times supported by state actors (Kloppenburg, 2004, p 80; Harwood, 2012). In Germany, these first professional breeders were typically feudal estate-owners in the plains of the north or smallholder farmers in the south (Harwood, 2012, pp 16–21; Brandl, 2017, pp 133–140). Different as they were in resources and philosophies, they set themselves apart from earlier breeders by carrying their seed to the market.

At first sight, seed is an ideal candidate for a commodity: it is usually small, durable and readily detaches from plants; it can be transported and, through reproduction, scaled up and down depending on demand. Compared to fresh vegetables or cut flowers, for example, seed is easily shipped and stored. But while it is essential for commodities to be able to travel to and from marketplaces unimpaired (Cronon, 1991, pp 230–247), this ability alone is not enough to guarantee that a good will function as a commodity. The fact that plants readily let go of their seeds does not mean they are going to end up in the right place.[11]

Transregional, anonymous seed markets were difficult to build at the turn of the 20th century. Why should farmers have abandoned their practice of saving seed from the harvest to instead buy from breeders they did not know? Fitzgerald (1990), Kloppenburg (2004), Moskowitz (2006; 2008) and Harwood (2012) go to great lengths to show how desire for and trust in new, improved seed had to be created among farmers by demonstrating its superiority to them before commercial plant breeding could commence. The historical problem was not limited to sceptical farmers, however. Supply was well ahead of demand, with commercial breeding in Germany beginning around 1850, while the seed market struggled to establish itself until at least the 1930s (Harwood, 2012, pp 34–37).[12] The problem, Harwood argues, was not that breeders failed to produce superior seed or that farmers were unaware of such successes. Rather, farmers, overwhelmed by the variety of choices offered by the seed market, failed to purchase the product they desired and, disappointed by commercial seeds, often resorted to their farm-bred seed again.

At least in part, the issue was one of a lack of information: farmers were buying seed, which, unlike their farm-saved seed, they were not familiar with. Quality assessment would have involved large-scale trials and comparisons between very different products. The obstacle faced by farmers was more than simply cognitive or calculative, however; it was also of a material nature. Once planted, even superior varieties often proved ill-suited for the local context of a farm. Plants that were well-adapted to the mild climate and ecology of the extensive northern German plains, for example, did not perform well when sown by peasants on the small hillside farms of southern Germany (Harwood, 2012, p 40). As its context changed, so did seed's quality change from superior to inferior. 'In today's language', Harwood (2012, p 40) writes, 'here was a straightforward case of "market failure"'. Although seed was readily alienated by breeders, the complementary operation by farmers – appropriation – failed. Even if purchasing seed from a distributor turned them into owners in a legal sense, commercial seed was by and large ill-designed to be re-contextualized in a practical and material way.

Farmers were not alone in struggling with commercial seed. The looming failure of the seed market – or at least its inability to reach a considerable part of its target audience – also represented a threat to breeders' businesses. From the breeders' point of view, the issue was not only that their seed all too often proved difficult to appropriate. In many cases, the opposite was true, too: commercial seed was appropriated too easily, which meant that they were alienating too much on the market. One of the constituents of seed that ended up in the wrong hands was its name. In 1920s France, the same wheat seed was sold under four different denominations (Sanderson, 2017, p 139). Around the same time, British commentators complained that several thousand different names were used for about 100 varieties, which in turn represented only 20 genuine types at best (Sherman, 2008, p 576). Meanwhile in Germany, out of 1,000 potato varieties in Germany only 130 were found to be genuinely distinct in 1930 (Harwood, 2012, p 84). Competitors either bought seed on the market, multiplied it and sold it under a different name, or used the name of a popular variety to sell their own inferior seeds to gullible farmers.

If the latter was simple fraud, the former practice demonstrated a particular weakness of the seed market. Resellers exploited one of the property scripts embedded in commercial seed to undermine their competition. This practice was an indirect result of pedigree selection, that is, the practice of inbreeding plants over several generations. Inbreeding allows breeders to measure, compare and replicate plants on a larger scale. It is a prerequisite for farmers to buy a standardized, reliable product that can be priced in accordance with its performance. In turn, however, it also enables others to circumvent the laborious breeding process (Sanderson, 2017, p 28). Through inbreeding, the script for replication is 'hard-wired' into an inbred seed: anyone with sufficient acreage could enter the fledgling industry by purchasing, sowing, harvesting and finally reselling seed – without having to go through the lengthy and cumbersome cycle of crossing and selecting. Whoever bought a bag of wheat also bought all the future generations contained in it, and there was nothing that prevented him to sell it to others.

The passage of property: making goods flow downstream

At first glance, farmers' and breeders' problems may seem unrelated: one concerns vertical relations (along the value chain) between producers and consumers of seed; the other horizontal relations between seed producers and their competitors. What connects both dimensions, however, is seed as a market good, linking them at the point of commercial passage. The very distinction between 'vertical' and 'horizontal' is at stake here, since the proliferation of seed through the market is what potentially enables consumers (buyers) to become producers (sellers). Commercial seed is thus not only confronted with market failure by flowing 'upstream', it also enables and produces this failure.

As Don Slater (2002, p 234) argues, 'markets are best defined in terms of a form of transaction rather than a specific mode of calculation: market transactions involve the alienation of goods in the form of property'. His statement might as well be turned around: markets are also about *appropriating* goods. Although market transactions are not the only way of transferring property, their mode of transfer is quite distinct and remarkable. The contrast between gifts and commodities is instructive here (Muniesa, 2008; Dobeson, Brill and Braun, 2023): the passage of gifts also involves the passage of a person tied to a thing; gift-giving thus makes personal ties, obligations and claims proliferate (Strathern, 1988). The same is true for several other economic forms and transactions, such as inheritance or renting: they rely on personal bonds that extend beyond the transaction itself (Braun, Brill and Dobeson, 2021).

In contrast, market goods are characterized by a particular form of passage. During a sales transaction, they are fully disentangled from their previous owners (Strathern, 1988, p 109; Perzanowski and Schultz, 2018, pp 15–16). This means that the seller completely cedes control of her property bundle to the buyer. In an ideal market, this happens simultaneously: the seller lets go of all the sticks in the same moment the buyer takes hold of them. If the spirit of the gift is characterized by a continued link between persons and things (Mauss, 2002), the spirit of the commodity rests in not being possessed by previous owners. Seller and buyer are 'quits', meaning there is no place for having second thoughts. Here property indeed resembles Blackstone's 'sole and despotic dominion' (Blackstone, 2016, p 1): notwithstanding our reservations against the simplicity of his idea, we would all be outraged if a supermarket reached out to us after our visit and told us what to do with our groceries or demand part of our purchase back. Having bought a commodity, we expect to be free from the claims of its previous owner. The downside of this arrangement is that sellers and buyers have only limited means to retroactively fix the passage of property if alienation and appropriation turn out to be misaligned.

For farmers this meant that they had paid for something that they could not fully make theirs. What unfolded before and after the market transaction was a set of scripts involving farmer, seeds, fields and local climate, in which the re-entanglement of commercial seed often proved considerably more difficult than that of farm-grown seed. Seed may have been stable genetically, but not ecologically, across environments. It was not just the farmer who needed to make a plant 'his'. The wider assemblage of farming routines and ecological processes had to accommodate commercial seed as well. From the farmers' point of view, appropriation through the market transaction was often insufficient: commercial seed lacked some of the scripts that would have allowed them to use it throughout the varied agricultural contexts of the country.

Breeders, on the other hand, suffered from a different problem. That farmers received too little did not mean that breeders were retaining too much. Quite to the contrary: the property issue for plant breeders was that they were alienating too much over the sales transaction. With inbred pure lines, they did not just hand over a technology for producing grain, they also sold a script for reproducing plants true to kind. Since the seed changed hands in a market transaction, there was nothing that barred buyers from using it for producing seed, rather than just grain. While this is true for almost any plant (Kloppenburg, 2004, p 37), the genetic stability of inbred varieties also allowed their seeds to re-enter the market as commodities, where they competed with the same varieties produced by the original breeder. By standardizing seed, breeders had created a commodity that circulated a little too well, undermining their own business model. If the fledgling market for seed was supposed to work, it needed to be fixed.

Remaking plants to fix the market

Unsurprisingly, actors on all sides sought to address the flaws of the seed market. In southern Germany, state institutions emerged in the interwar years, which developed, tested and recommended plants that were adapted to local environmental conditions, distributing them to farmers for free (Harwood, 2012, pp 57–75). Meanwhile, commercial breeders tried to tackle the same problem by cooperating on the large-scale testing of new seed. At the same time, they unsuccessfully attempted to combat seed resale with patent and trademark protection. Commercial and public breeders fiercely debated whether the choice between an increasing number of different seed products was of use to the farmer, whether private testing was biased or if public breeding represented a distortion of the market (Harwood, 2012, pp 79–80).

Trademark protection proved too loose: it would stop third parties from selling seed under a particular denomination, but not from brown-bagging it as a different one. Names, even when protected, detached from bags of seed all too readily while new ones could be stuck to the same bags with ease. Patents, in turn, were too strict: while preventing others from reproducing a certain variety, they would, crucially, also bar breeders from using their competitors' seeds as parents for new varieties (Snell, 1939; Bent et al, 1987, p 46), a common and essential practice for German seed companies to this day. Moreover, patent offices rejected applications for plants in the interwar years because breeders could neither explain how they 'invented' them nor detail how others could independently arrive at the same plants (Fowler, 2000; Pottage and Sherman, 2010, pp 153–182). The question of quality and reliability, most urgent for the farmers, was finally tackled through market admission trials following the French model (Bonneuil and Thomas, 2009, pp 60–61) as part of the National Socialist regime's attempts to strengthen smallholder agriculture (Harwood, 2012, pp 98–106; Saraiva, 2016, pp 85–87). Similar steps were taken elsewhere in Europe in the interwar years (Heitz, 1991, p 25) but fell short of protecting breeders' interests.

The market would not truly be fixed until after World War II, however. The solution that emerged in post-war Germany (and elsewhere in Western Europe) did not simply redistribute responsibilities between private and public breeding and testing programmes. More importantly, it redefined seed biologically, legally and economically. In 1953, Germany introduced a national Seed Act (*Saatgutgesetz*), which prescribed the material characteristics necessary for commercial seed.[13] In order to receive approval by the newly created Federal Plant Variety Office (*Bundessortenamt*), the law required seed to be 'distinct' (morphologically distinguishable from other varieties), 'stable' (able to pass on its genotype unchanged) and of 'agronomic value', that is, increasing 'yields of a smaller or larger region' (Gesetz über Sortenschutz und Saatgut von Kulturpflanzen (Saatgutgesetz), § 2, nr 4).

These provisions were not limited to a quality guarantee to farmers. They also had profound implications for the material form of seed that could enter into economic circulation. Of central importance was the concept of the variety (*Sorte*): while it had been used long before the law was enacted, it now received a precise meaning beyond a taxonomic rank below the species (Sherman, 2008), referring to seed characterized by a particular set of features. The distinctiveness requirement barred all varieties from the market that could not be shown to be sufficiently different from existing ones. Phenotypic stability was already a prerequisite for many breeders for selling the same variety over subsequent years, but it also became crucial for an emerging public assessment apparatus coordinated by the variety office. Seed was only admitted if breeders could guarantee that it was identical to the one tested by the research stations. One result of this new accounting apparatus was a 'national environment', which breeders had to take into account when developing new seed.[14] Together with the emergence of a new regime of chemically supplemented agriculture (Uekötter, 2012), which effectively offset local ecological differences, this guaranteed that varieties that made it to the market could be planted anywhere in Germany.[15]

The variety office developed a classification system of 'register characteristics' for distinguishing between individual varieties and for connecting one specific phenotype to one specific name, which enabled breeders, farmers and authorities to settle the identity of a variety. Names were fixed to a particular variety and could not be used for any other seed. In turn, breeders were prohibited from selling their seed under any other name than the one under which it had been admitted. This purged the market of generics, reduced options, created unambiguous references and identities, or, speaking with Cochoy (2007), tended the garden of choices in seed retailing.

The emerging assemblage of product standards, chemical technologies, state-approved information and certification, as well as a slimmed-down market allowed farmers to appropriate seed much more easily. While there were still considerable differences between plant varieties, farmers could rest assured that new ones improved upon existing ones – at least when measured against the standards of the new national assessment framework. Fixing names to plant varieties prevented both from changing their identity when travelling across the German seed market. All of these measures, however, only solved the farmers' problems. They did not do much to address breeders' concerns about the misappropriation of plant varieties by buyers-turned-competitors. Worse even, while the Seed Act restricted breeders' ability to alienate seed on the market, increased standardization and stability made it even easier to reproduce varieties.

For this reason, the Seed Act introduced a new property regime, exclusively limited to the domain of plant breeding: PVP. Where both trademark protection (which only protected product names) and patents (which only protected a technical principle) had failed in the interwar period (Fowler, 2000; Pottage and Sherman, 2010; Sanderson, 2017, pp 21–44), PVP included elements of both, aiming to protect a commodified market product. Fusing name and form in one and the same property object, it gave the original breeder of a variety the exclusive rights to take it to the market.

Reporting on behalf the working committee for the Seed Act, MP Robert Dannemann (Free Democratic Party) noted:

> Deliberately, the Seed Act refrains from the following provisions: 1. Rights to prohibit or exclude regarding uses not linked to the production and commercialization of seed. 2. Rights to prohibit or exclude pertaining to seed production and distribution insofar as these acts are not performed with the purpose of commercial distribution of seed. (Bundestag Drucksache Nr 01/4339, p 5, translated)[16]

He further stressed that 'according to patent law, anyone drawing from the patented invention of another (dependent inventor) requires permission of the first inventor. In contrast, according to … the Seed Act, the "dependent breeder" does not require the first breeder's permission for using seed of the dependent variety'.[17] In other words, breeders were exempt from other breeders' property claims so that one could not exert power over the other through her property object. 'Plant variety protection only covers seed produced for the purpose of commercial purposes ...', Dannemann emphasized, 'seed produced by the farmer for his own operation is thus free' (Bundestag Drucksache Nr 4339, p 5). The proposed legal script was thus intended to interrupt one particular movement of seed after sale – its return to the market. No restrictions were put on its circulation back into breeders' nurseries or on the farm, although the question of where to draw the line between breeding and agronomy was a controversial one.

After a heated debate in parliament, lawmakers eventually agreed that 'exchange across the fence' of saved seed was where breeders' rights should end.[18] As a consequence, PVP became a much slimmer instrument than either trademark or patent protection, a fact for which it is still widely praised to this day (Kloppenburg, 2014; Brandl, 2017). In the years following the Seed Act, a group of 12 Western European countries worked towards harmonizing their national PVP laws (Bent et al, 1987, pp 40–42). These talks eventually resulted in the UPOV convention of 1961, which would be expanded in the decades to follow (Sanderson, 2017) and cemented the ambivalent relationship between commodification and de-commodification of seed under PVP. If patent law knows a 'patent bargain', which grants property protection in exchange for disclosing technological knowledge (Biagioli, 2019, p 21), UPOV introduced a 'variety bargain', in which breeders received extended property rights in exchange for making their seed distinct, uniform and stable – that is, materially commodifying it.

Where the 1951 Seed Act conflated the horizontal (competition) and vertical (consumer protection) dimensions of seed markets, UPOV again separated them, forcing Germany to create two distinct laws in 1963: the Variety Protection Act (*Sortenschutzgesetz*) and the Seed Marketing Act (*Saatgutverkehrsgesetz*). UPOV member states were required to harmonize intellectual property laws to ensure fair competition but were otherwise free to set different standards for marketing seed. Nevertheless, the distinctness, uniformity and stability (DUS) criteria still bore testimony to the implicit assumption that only seed fit for industrial agriculture could be protected (Kloppenburg, 2004, p 150).

Giving to or taking from the farmer? Seed as a quasi-commodity

Two aspects of the PVP story are paradoxical. Although the parties involved were aware that too much was being alienated on the seed market, the post-war answer to the problem was not to de-propertize seed, for example by completely entrusting it to the public domain, nor to commodify it altogether. True, seed was further commodified materially through PVP, but at the same time, it was legally de-commodified: farmers no longer received the full bundle of property sticks. In adding another layer of property on the possession of seed, PVP represents an instance of propertization. Rather than disrupting the market further, however, this re-propertization and de-commodification made seed a better, not a worse market good.

Here we encounter a second paradox: it cannot be overlooked that seed bears an uncanny resemblance to more recent property objects that come with 'strings attached' – digital music and books, paid content, technological use restrictions (Doctorow, 2008, pp 3–28). Even beyond the point of sale, these objects still belong to someone else, limiting the freedoms of use buyers would otherwise enjoy while keeping the producers in the picture (Perzanowski and Schultz, 2018). In the same vein, PVP stops seed from full alienation and farmers from using it in certain ways. Breeders retain a backdoor through which their claims are reactivated once their seed reappears at the market. This is what attracted Kloppenburg's (2004, p 151) suspicion, who saw the corresponding US Plant Variety Protection Act of 1970 as the harbinger of patents on seeds and as an instrument for dispossessing farmers. Yet it appears that in the 1950s and 1960s farmers did not take issue with the introduction of PVP, even if on paper they ceased to be the absolute sovereigns over 'their' seed: for the next three decades, plant variety protection by and large remained an uncontroversial topic.[19] The continued power of seed producers over seed and farmers would become a point of contention with the advent of genetically modified seed (Müller, 2006; 2015; Schubert, Böschen and Gill, 2011).

How was it possible to reconcile de-commodification on a legal level with the general commodification of seed after World War II? Why did breeders' persistent property ties to their seed not enter into open contradiction with the otherwise so central principle of alienation? The answer lies in the larger context of economic relations between breeders and farmers as well as the latter's understanding of their own profession. As far as farmers were concerned, they primarily bought seed to sow, harvest and then sell it to processors. While some farmers bought seed, reproduced and then sold it, this was not a traditional farming practice, unlike reusing seed for one's own farming purposes. The wider bundle of scripts was thus barely affected by the redistribution of property through PVP – at least as far as it involved traditional breeders and farmers. The emerging group of farmers-turned-resellers, however, was banished from the market by the Seed Act. PVP firmly drew boundaries around two professions that were not supposed to mix, while simultaneously defining the trajectory for the social life (Appadurai, 1986; Tsing, 2013) of seed as a market object.

The economic landscape this object was embedded in was one in which both breeders and farmers played an important role in the war against hunger (Harwood, 2012; Uekötter, 2012). Both groups were politically and economically important. By carefully balancing the bundle of rights and practices attached to seed between them, post-war legislators in Germany and elsewhere in Europe managed to find an arrangement that allowed both farmers and breeders to prosper. In this arrangement, the practice and right of seed reuse for farming purposes were left with the farmers. Despite increasingly buying their seed at the market, they could thus still *act like* sovereign owners of their purchase. In PVP jurisprudence, this freedom would come to be called the 'farmers' privilege' (Sanderson, 2017, p 189). In turn, breeders could use competitors' varieties for breeding – the 'breeders' privilege'– and rest assured that they were the only ones who could produce seed for the market.

To loosely borrow from Michel Serres (2007, pp 224–234), PVP seed was a 'quasi-commodity' – one whose nature as an entangled object was irrelevant for most farmers because it only affected those who left their role as customers to become competitors.[20] Although seed's materiality is unique, the parallels between PVP and other forms of 'intellectual' property are remarkable. Property rights are not attached to an idea or abstraction, even if intellectual property law entertains this fiction (Peukert, 2021).[21] Instead, they are tied to the very same things that we consider 'concrete' or 'embodiments' of such abstract entities. When we buy, say, a Samsung fridge or Nike sneakers, we rarely consider the possibility that these things still belong to their producers in one way or another. But as soon as we start using them to build a fridge similar to Samsung's or prominently wear them in a video that does not align with Nike's public image, we will hear from the companies' legal department and the entangled ownership of 'our' possessions will become apparent.

If we can act as if our fridges or shoes are ours and ours only, it is because it rarely occurs to us to use them in the few ways that are covered by utility and design patents, trademarks or copyright. The vast majority of our uses is indeed free from others' property claims once we purchase commodities – to use the technical term, their property rights are 'exhausted' over the sale (Bently et al, 2022, pp 14, 721). Yet this exhaustion is a fiction that can only be upheld if the gulf between using our possessions one way (as consumers) but not another (as competitors) is sufficiently large. That we do not experience our fridges, cars, clothes or balcony plants as co-owned, is usually because we lack the material means to do so in the first place. The full alienation of most goods is thus a fiction, albeit a less obvious one than the alienation of seed: after all, these things do not disappear from this world over the sales transaction but continue to circulate in it. Intellectual property seeks to make sure that this circulation will not cross the market in ways that would undermine the continuing interests of producers and sellers. For this reason, intellectual property needs to make a distinction between producers and consumer, between competition and intended use.

With seeds, this was the case back when PVP was conceived: crop farming did not rely on the mass-production and sale of commercial seed, hence farmers had little to quarrel over with breeders. Drawing the boundary between nursery and farm at the market, PVP legislation thus did not only give shape to commercial seed. It also cemented a particular distinction between breeding and farming, two practices that only a century earlier had been synonymous. In the same breath, it cleared the market of resellers, brown-baggers and commercial reproducers whose activities and products could no longer be accommodated in the seed market since they transgressed the legal spaces of nursery and farm.

The Solomonic solution of PVP consisted in rendering unto breeders the practices that were the breeders' and to farmers those that were the farmers': breeders produced seed, farmers produced grain. As long as seed continued to flow down the value chain, it would not reappear on the market. Farmers could stick to their customs without being affected by the property claims that still rested in their seed. All the new property regime asked of them to not turn this seed back into a market good. With PVP, German post-war seed thus managed to be the servant of two masters, rather than following the either-or rationale of conventional commodities. Although legally, practically and materially very different from classical market goods, seed emulated their trajectory and successfully extended the reach of market economics to the realm of plant breeding.

Conclusion

After an eternity of waiting, the door to the hearing room opens and Wolfhard, Sabine, Hubert and I are asked to return to receive the verdict. The day turns out as one of mixed success: out of five varieties, two receive admission to the market while three are rejected. This means that the company will have two new products out on a very dynamic market, but it also means that their property in three other products will remain more exclusive than they would have wished. For these three varieties, all the years of crossing, selecting, evaluating and going through trials have been in vain. For the two varieties that have been admitted, Hubert's company will receive between 14 and 25 per cent of the retail price, depending on the market segment and the price of competing varieties (Braun, 2020).[22] These revenues are what the company can use to develop new varieties (and to cover the losses from failed ones).

Economists often stress the importance of property for the market and the economy. Without property rights, they argue, people would simply take things, rather than pay for them: hence why we need property instruments such as copyright or patents, which force people to go through the market when they want to acquire books or inventions (Machlup, 1958, p 3; Landes and Posner, 2003, p 12).[23] The sole emphasis put on the exclusivity of property is often misplaced when it comes to the market, as Posner (1986, p 31) recognizes: breeders do not want to exclude farmers from their varieties. Much to the contrary, they want to hand their varieties over to them – albeit in a very defined and controlled way. To make sense on both sides of the market transaction, however, alienation needs to be mirrored by appropriation on the farmers' side. Plant varieties are transferable as property, which is a more-than-legal matter: breeders need to be able to alienate seed, but farmers must also be able to appropriate it. Appropriation is achieved by producing seed which does not require a specific knowledge, ecosystem or farming practice – but it also relies on a complementary system of agriculture which can level such differences by additional chemical and mechanical inputs. Agronomic, economic and legal friction is thus reduced; the flow of seed through market transactions is facilitated. As a result, property on the market represents a delicate balance of scripts for alienation, appropriation and exclusion.

Both sides of this 'double movement' are equally important. If farmers cannot grow Hubert's seed, he cannot sell it, but if he does not receive permission to commercialize his varieties, farmers will be unable to purchase them. Alienation poses a second, material challenge, however. If Hubert sold his customers too much or if they appropriated more than they paid for, Hubert would have to quit his job sooner or later. To square this circle, PVP legislation did not shut the world out. Rather, what it excluded was one single yet crucial right – to commercialize seed – from the bundle to leave it the breeder's hands. Down- and upstream the value chain, however, in variety development and in the farmer's field, varieties remained free from the breeder's property claims. For this reason, property in the market is better understood as a way of making things move in a particular fashion, as channelling the passage of property in goods, rather than a divide between haves and have-nots.

If the relationship between breeders and their varieties is to be called one of private property, it is only in the context of the market, whose order and functioning the exclusive right to commercialize is meant to ensure. During this brief moment of their social life, legally protected varieties behave like the exclusive good that Posner (1986, pp 31–39) postulates. PVP is thus minimally invasive, enclosing only a particular practice pertaining to seed. It is for this reason that I call plant varieties 'quasi-commodities' here: in their characteristics, they largely follow the model of consumption goods (Graeber, 2011), except that their material exhaustion is not obligatory. To compensate for this deficit, the market transaction must be stopped from fully alienating the breeder's property, giving her a legal leash to pull back her property if it goes astray.

Despite the imperfect commodity character of seed, its reframing as property has been successful in the sense that it has contributed to the emergence and long-term success of competitive markets for seed. Together with the steady proliferation of ever-new and improved plant varieties, it is an achievement private plant breeders take great pride in. 'Without property', Hubert underscores, 'there is no breeding progress'. It is of course not the only conceivable way in which farmers can be supplied with suitable varieties. The US (Kloppenburg, 2004, pp 39–49; Brandl, 2017), Canada (Müller, 2006) and Australia (Head, Atchison and Gates, 2012) have historically found ways to entrust plant breeding and variety development to public research in many crop species, turning seed into a tax-funded public good.[24] As an alternative to industrial agriculture and PVP, organic agriculture and farmers' seeds activists have developed networks and practices for seed sharing which do not revolve around alienation, exclusion, market-compatibility or 'authorizing' (Gill et al, 2012) a particular breeder, but rather around sharing and contributing and how to make these practices proliferate (Aistara, 2014; Demeulenaere, 2014; Kotschi and Horneburg, 2018). Certainly, these alternative approaches come with different implications for breeding practices, economic growth and yield increase (Brandl et al, 2014; Brandl, 2017; Brandl and Glenna, 2017). Still, the continued existence of such alternative ways of producing and provisioning seed means that they are not entirely unfeasible or unsustainable and that there are different ways for making breeders, farmers and seed meet. Property is one solution to a problem that can be framed and answered in very different ways.

The important lesson from the history of the seed market and the introduction of PVP, however, is that markets can indeed be fixed with property, and to the satisfaction of the stakeholders they are meant to serve. What is remarkable in this case is that the same property instruments that distort, circumvent or undermine consumer markets today (Perzanowski and Schultz, 2018) stabilized the seed market 70 years ago: propertization had not yet run out of steam. Seed became a good with strings attached but still successfully mimicked an ordinary tangible property object. This paradoxical arrangement was possible because breeding and farming took place in separate social realms. The emerging space where these realms started to overlap – large-scale commercial reselling – was closed by PVP legislation before it could fully materialize.

The Seed Act and later PVP legislation not only relied on this separation. They were also based on a number of explicit and tacit assumptions: the reliance of food security on domestic farming; the autonomous smallholder as a sovereign of the farm; the sufficiency of market revenues to support not only variety development but commercial breeding in general; the free flow of germplasm and genetics into the nurseries of breeders. As such, PVP is a child of the mid-20th century, a time where many traditional models in agriculture started to change but still informed agricultural policy and reasoning. In the decades to follow, farm sizes would increase in industrialized countries and farmers would specialize, abandoning the model of the farm as a comprehensive agricultural operation that produced a varied assortment of foods for the market (Kramer, 1980; Fitzgerald, 2003; Uekötter, 2012, pp 339–379). Meanwhile, breeding has also undergone slow transformations. If plant breeders could expect reliable incomes from their admitted varieties for the full PVP term in the 1960s, the commercial life of varieties has drastically shortened today: Hubert can be lucky if his varieties fetch the bulk of their revenues in the first two to three years before dropping out of a highly competitive market (Pallauf, 2018).

For seed companies, this also means that they can rely less and less on the money from seed sales. At the same time, developing new traits has become more cumbersome and expensive. Genes coding for resistances or chemical components of plants are often derived from wild crop relatives or landraces whose form is incompatible with the industrial, chemically supplied agriculture PVP takes as its model (Bertacchini, 2008; Fullilove, 2017; Dutfield, 2018). The more the progress in the development of standardized commercial varieties advances, the longer it takes to channel traits from such unruly material into commercial plant varieties – the appropriation of traits has become more expensive and labour-intense. PVP, however, is silent on this matter: it assumes that breeders will invest in the development and improvement of plant properties out of their own interest. With ever-rising costs of trait development, this assumption becomes increasingly tenuous. Hubert admits that there are colleagues who deem his investments in resistance breeding a waste of money. At the same time, they openly joke about using his varieties as parent as soon as they hit the market under the breeders' privilege. This is a practice that is not only allowed by PVP but actively encouraged as part of the breeders' privilege, since it facilitates the spread of beneficial traits through the commercial gene pool.

Breeders, including Hubert, nevertheless defend their privilege, arguing that it is the foundation their industry is built on. Still, they are becoming increasingly conscious of the dwindling profitability of variety development and the rising costs of trait breeding. Seed companies and their lobby have reacted to this problem in different ways. As we will see in Chapter 6, one strategy consists in renegotiation the post-war consensus between farmers and breeders in the latter's favour. By shrinking the farmers' privilege, the hope was to recoup more of the investments that plant breeders need to make to survive. As a result, seed is showing its true nature as an entangled object: over the various revisions of UPOV and PVP, it has become subject to new claims by breeders and increasingly politicized among farmers. Another strategy has been to look for additional revenues not down- but upstream the seed value chain: why should breeders not have to pay if they want to use a resistance gene, leaf colour or plant compound a competitor has integrated into her commercial material at great expense? Although Hubert and almost all of his colleagues object to this idea, it has gained increasing ground in the European seed sector, materializing in the European Patent Office's decision in *Broccoli/Tomatoes II* to grant patents on traits.

4

Re-inventing Plants

In late March of 2015, while I am out doing fieldwork at Hubert's breeding station, news arrives from Munich. The EPO has finally reached a verdict on *Broccoli/Tomatoes II*. To the great protest of plant breeders, seed activists and legal scholars, the EPO's Enlarged Board of Appeal has ruled that plants, even if they had not been subjected to biotechnology, are patentable. The board found no reason to deny patents on conventionally bred plants, thus affirming the approval of two controversial patent applications on a self-drying tomato and a broccoli (Braun, 2023). Clearly oblivious of Rousseau's (2011, p 70) reminder that the fruit of the earth belong to all, the patent office has ceded them to the Israeli Ministry of Agriculture (*Tomatoes*) and a British biotech company, Plant Bioscience Ltd (*Broccoli*).

On the face of it, there is indeed something scandalous about this ruling. The scandal is not just about private property as a robbery of the public (Proudhon, 1994). It is also about an understanding of plants as inventions, as the products of human ingenuity and agency, something that is at odds with the millions of years of evolutionary inventiveness of plants which owe very little to humankind. Indeed, we may ask ourselves how plants have become the object of patent law: is the latter not about invention and technology? And are conventionally bred plants, produced by the biological mechanism of cross-fertilization and the simple technique of selection, not the very opposite of these concepts? How did we get from Rousseau's statement to the EPO's position on plants in the *Broccoli/Tomatoes* case?

The aim of this chapter is to understand how the Enlarged Board of Appeal could, in spite of all objections and criticisms, declare ordinary plants patentable technologies. The ruling marks yet another advance of propertization, another step in the continuing expansion of patent law (Eisenberg, 2006). Perhaps, however, we can learn more from this event by understanding it as a coincidence or accident of, rather than just the latest scandal of propertization. After all, most of the stakeholders and observers agreed that the EPO had made a mistake in *Broccoli/Tomatoes II*. Plant breeders' lobby organizations issued statement in which they objected to the decision (BDP, 2015; Euroseeds, 2019). Axel Metzger, pre-eminent commentator of intellectual property law in plants, saw '[l]egal [p]ositivists at [w]ork' behind the ruling (Metzger, 2016). Even the Swiss biotech giant Syngenta, not otherwise known as a patent sceptic, had opposed the *Broccoli* patent and taken the case to the highest appellate body of the EPO in the first place. If all these parties had voiced their objections, then why did the Enlarged Board of Appeal press ahead with its ruling?

Siegrist defines propertization as 'intentional strategies developed by identifiable actors in specific contexts and pursued with purpose' (Siegrist, 2005, p 104, my translation), an interpretation that fits the history of PVP in the post-war seed market as retold in Chapter 3. Here, legislators and plant breeders turned seed into a particular form of property – or rather added another layer of property to it – to make it work as a market good. Such a reading of propertization gives an air of instrumentality to property: it is used to pursue someone's particular interest, likely at the expense of others. In such a context, the question of '*cui bono?*' (to whose benefit?) readily suggests itself (Star, 1995). In the preceding chapter, I have answered it with 'both sides', although the usual reply would likely be more-one sided, followed by the related question of '*cui malo?*', to whose detriment? To ask in such a way is part of a critical, political-economic toolkit that understands economic and legal processes as political struggles in which the powerful secure their interests at the expense of the weak or society as a whole. Kloppenburg's (2004) account of propertization in plant breeding can be read along these lines, as the successive extension of seed producers' property and the increasing dependence of farmers on them. To understand the *Broccoli/ Tomatoes II* case in this light, we would thus have to ask who benefited from the EPO's decision and inquire how they pursued their interests before and during the legal process.

However, Siegrist also gives us a second interpretation of propertization to complement his first definition: the 'self-driven processes that play out, as it were, behind the backs of the actors' (Siegrist, 2005, p 104, my translation). In this wording, actors are no longer in the driver's seat. Instead, propertization is something that happens to them as much as it happens to the things being propertized.[1] In this chapter, I want to follow this second definition of propertization to explain how the usual suspects – plant breeders, seed multinationals, lawyers, perhaps even the European Patent Office – found themselves in a situation none of them had planned, orchestrated or pushed for. Propertization, I will argue here, has a dynamic of its own, which cannot be reduced to the interests of the powerful, the logic of capitalism or any other overarching force. If we want to understand why property is becoming increasingly problematic, it is not enough to think about it as an instrument of coercive power (Pistor, 2019; Robé, 2020). Instead, we need to think about the power inherent in the idea of property itself. To this end, I will attempt here to sketch plants' transformation into patentable objects as a series of material, epistemic and legal changes that in hindsight form a trajectory but cannot be boiled down to power struggles. Over the course of this history, plants only seemingly remained the same: several transmutations over the last 100 years made it possible to think of them as inventions. Similarly, what we conceive of as a patent has shifted over the same time. Patents today are not the same thing they were a century ago, meaning that the categories of property themselves are fluid.

The other plant intellectual property

There are many possible beginnings for patented plants in Europe and the *Broccoli/Tomatoes* saga. Perhaps not readily obvious, one lies in the United States. Although commercial plant breeding suffered from similar problems on both sides of the Atlantic, the legal history of plant innovations took diverging paths in Europe and the US. As we have seen, European countries struggled to accommodate seed in their existing IP frameworks, particularly patent law. They finally agreed on a *sui generis* right for plant innovations which, as one of its results, led to standardized, market-compatible plants. The concern in the US, in contrast, was more with the nursery than with the market. Here, much of the discussion revolved around the protection of novel plant traits, rather than commercial varieties. Despite their differences, however, both legal systems had acknowledged similar ontological – and therefore legal – difficulties with respect to plants and patents (Fowler, 2000; Kevles, 2007; 2011; Sanderson, 2017, pp 81f). Applied to the field of plant breeding, central terminology of patent law did not make sense: if a breeder had discovered a peculiar sport in his nursery, was he the inventor of this plant? Was this plant novel in the first place, and if so, how was this novelty to be proved? How should the essence of a novel plant be documented in a patent description, especially if a patent should enable others to understand and repeat the invention? Was it possible, in the first place, to reproduce such a sport? If so, *who* would reproduce it – a breeder or the plant itself?

Arguably, such questions were more pronounced in the US discourse, which circled around the breeder as a sort of creator of novel plants, whereas in Europe they mixed with questions of market order and transparency. Framing the problem as one of calculability, agronomic adoption of plant seed and its downstream movement in the value chain, Europe gave a decisive push towards a market solution that would turn seed into a commodity. PVP justified breeders' property in varieties with the money and labour they had invested in them, thereby answering what remained an open question in patent law, namely if the breeder's intellectual contribution to his plants amounted to something like invention. With the legal category of the variety, however, PVP also set rather narrow limits for the kinds of plant innovations that could be appropriated with new IP instruments: sports, mutations or buds certainly did not qualify for PVP. In comparison, the US Plant Patent Act of 1930 stuck more closely to the fundamental openness or indeterminacy of patent law. It did not dictate what form a plant had to take in order to be appropriated; instead, it sought to make room for as many novel and unexpected traits as possible. While the scope of protection was limited to asexually reproduced plants (which were considered 'true to form'), there was no comparable limitation concerning the form of a novel plant – it could be any 'new variety of plant, including cultivated sports, mutants, hybrids and newly found seedlings'.[2]

Pottage and Sherman (2010) point out that the crucial divergence between the mode of innovation in early 20th-century plant breeding and industrial production can be found in the relationship of what the latter considered forms or originals on the one hand and exemplars or copies on the other. One of the key features of 'mechanical' patent law, according to Pottage and Sherman, was that it conceived of inventions in terms of 'originals' that could be copied.[3] Originally, there was an original, a form – printing letters, cast forms – that could impose the invention onto a piece of matter such as paper or metal. As Pottage and Sherman note, '[t]he (invented) original became the template from which a succession of identical exemplars was produced, economic value being concentrated in the (patented) idea rather than its material embodiments' (Pottage and Sherman, 2010, p 42).[4] While in the beginning of patent law, such originals were of a similar nature as the exemplars, taking the forms of prototypes or miniature models, late 19th-century patent law abandoned them in favour of technical doctrines exemplified in text and drawings. The US patent act of 1870 dropped the requirement of inventors having to submit a working model with their patent applications, demanding a textual specification instead (Pottage and Sherman, 2010, pp 94, 142–144).

Regardless of how an innovation had actually been produced up to the patent application, this relationship between an intangible form and a tangible exemplification, in which only the former is directly relevant for the grant of a patent, assumes that there is a vector, which runs from idea to realization. This is commonly considered the essence of inventing: a person conceives of an abstract technical principle and then turns it into an apparatus or system, which is guided by this principle and will apply it in the same breath. While inventing may look very different from this idealized version in practice (Collins, 1974; Misa, 1992; Pickering, 1995; den Hond, 1998), it is important that we can imagine and understand the legal fiction (Fuller, 1970) of invention this way: as a straight vector running from conception to exemplification. Invention is not mere discovery of a thing already 'out there', which can be imagined to exist without the additional input of an inventor. The concept instead implies a certain form of creative agency as the cause behind the essence of the property object claimed. Furthermore, to invent also means to come up with an idea first, not to receive it from someone – or something – else (Sanderson, 2017, p 81–84).

In the case of plants, this concept of invention proved difficult to apply, as Pottage and Sherman note, because innovation did not follow the succession of conception, development and application:

> Classically, the inventor was supposed to have evolved the inventive concept to the point of coherence and practicability at which it could be rendered as a (textual) genetic template – a recipe, or a mould from which copies could be produced. Reversing the direction of this process, plant breeders argued that new plant varieties were invented by identifying unique mutations, stabilizing them, and turning them into templates from which a succession of copies could be made. Invention was inductive rather than originating; it consisted in retroactively turning something found into a reproducible form, rather than originating a new (and reproducible) form. (Pottage and Sherman, 2010, p 153)

Seeking to protect their work with patents, plant breeders argued that even if it did not constitute an act of invention, at least the categories of the patent applied: there was an initial exemplar which, although discovered instead of invented, was turned into a form by inbreeding and multiplication (Pottage and Sherman, 2010, pp 163–165).

Unlike European PVP, which came to acknowledge and protect work and investments in plant breeding but not inventing as the appropriating act, US legislation did not discard the patent option. However, it could not simply gloss over the ontological differences between Thomas Edison's light bulbs and Luther Burbank's *Gladiolus* bulbs. As a result, plant patents followed breeders' line of argument, but perhaps more closely than the latter would have liked. Asexually reproduced plants could be protected, but sexually reproduced ones could not. To prevent a conflict between patent rights and the practices of farmers, tubers – which included potatoes – were further excluded from eligibility. This meant that all species of agricultural importance at that time fell outside the scope of the novel patents. Uncertain of their ontological status, the 1930 Plant Patent Act spoke of breeders as '[w]hoever invents or discovers and asexually reproduces any distinct and new variety of plant' (35 US Code § 161), leaving open if there was a difference between discovery and invention in plant breeding and, if so, what it amounted to. Most notably, however, it articulated property through the original act of invention, discovery and reproduction and not, like PVP, around the final product. What qualified a plant innovation for US plant patent protection was not utility in patent law's sense, usefulness or practical application, but simply distinctiveness (Kloppenburg, 2004, p 133; Kevles, 2011, p 264). Plants that could have proven most practically or industrially useful had been exempt by the restriction to asexual reproduction.

Taken together, these limitations effectively made the 1930 act an IP right for horticuturalists. They followed the model of the curiosities Burbank had come across and developed in nurseries (Fowler, 2000; Pottage and Sherman, 2011, pp 156f),[5] who thought of breeding in terms of spectacular traits like disease resistance, spikelessness or novel colour – in contrast to commercial field crop breeders whose product was pre-packaged, ready-to-use seed. Unlike cereal and pulse breeders, who had to rely on inbreeding to achieve somewhat stable transmission of traits from plant to plant, horticulturalists (i.e., fruit tree and ornamental breeders) could exploit asexual reproduction which would yield genetically much more stable plants.

While Kloppenburg (2004, p 132) dismisses the Plant Patent Act as a (failed) lobbying attempt for a monopoly over food crops, Cary Fowler thus characterizes it as an instrument for resolving conflicts between competing horticultural breeders:

> Unlike the seedsman, the nurseryman's biggest competition came not from the farmer, but from the competing nurseryman who could legally buy superior or unique stock and set up a large-scale operation to multiply and sell it. … Since fruit trees remain productive for decades, a lost sale was an opportunity lost for a long time. L.H. Bailey's list of plants introduced in 1893 gives some evidence of the trouble certain nurserymen were beginning to have. Each of the corn varieties introduced that year was offered by a single company, whereas most of the rose varieties were offered by a multiple 'dealers'. (Fowler, 2000, p 627)

Therefore, the US Plant Patent Act was directed less at an extraction of surplus from farmers through artificial monopolies (Kloppenburg 2004, pp 19–49) but, like PVP, targeted 'unfair competition' at a horizontal level: that is, the practice of asexually reproducing plants without investing into the examination and selection of plants. It did not grant breeders any excessive claims to methods, which were considered even less patentable than plants (Sanderson 2017, p 91), only to products. As a legal fence, it was called into existence because material fences and various other scripts for enclosing property had proven ineffective.[6]

By following the model of horticulture and its comparatively easy transmission and combination of genes via cloning and grafting, the US Plant Patent Act left an open space that in Europe was closed by PVP. For decades to come, field crop breeders would remain without a dedicated property right, instead relying on revenues from hybrid maize, plants developed by public breeding and business relations (Charles, 2001, pp 112–117; Brandl, 2017, p 123). The European framework, with agricultural crops in mind, strictly defined what an innovation had to look like, whereas the US act did not carry any comparable division beyond the requirement of distinctness. The question of what plant innovation was to be, both as a practice and as a thing, was thereby answered in Europe: it was not based on ingenuity, but hard work; it would yield a materially commodified product; and it would be in the overall service of farmers and agriculture.

US law, however, was less clear about the ontology of plants, their development by breeders or their commercial value. 'Variety', in its sense, could mean anything from one unusually coloured flower to an apple tree planted thousandfold in orchards between New York City and Los Angeles. In its mixing of the personae of the inventor and the discoverer, it furthermore blurred the boundaries of previously established boundaries between technology and science in patent law (see Pottage and Sherman, 2011, pp 159–166). Finally, if not in extent then certainly by its name, the plant patent was a patent. It was modelled on and modified from the more familiar utility patent to fit the domain of plants, but it strengthened the idea that the concept of a patent, in one form or another, could be applied to living things.

The US Plant Patent Act may be seen as the failure of agricultural breeders to obtain property rights comparable to those of their European counterparts a few decades later, and certainly many of its assumptions proved to be impractical once it came to lawsuits. Claimants often could not prove infringement, inventorship reemerged as a shaky category and the relationship between idea and embodiment remained an open question in plant breeding (Pottage and Sherman, 2011, pp 166–180; Fowler, 1994, p 93). At the same time, by answering the question of applicability of classical patents to plants in the negative, it set the realm of plants apart from chemistry and mechanical industry. Nevertheless, in carving out a first legal foothold – brittle as it may have been – the Plant Patent Act formed the basis for a future expansion of property that vastly exceeded PVP.

Ambiguous genes

One of the most notable features of the Plant Patent Act, compared to plant variety protection, was that it offered the possibility to legally appropriate genes or, more correctly, traits. Although Mendelism was undergoing its renaissance at the time of the conception of the act, genetic theory was far from unchallenged or universally accepted by breeders (Fowler, 2000; Bonneuil and Thomas, 2009, pp 38–57).[7] Nonetheless, while early 20th-century breeders may have debated the origin of plant traits, their debates revolved precisely around such individual characteristics. It therefore does not surprise that they were given according space in the Plant Patent Act. Genes, however, meant something else in 1930 than today. While traits were easily visible on an individual plant, genes were characterized as that which was passed on between generations and caused traits. In this sense, novel and unusual traits reproduced by asexual means were not genes, as they did not bridge generations but were taken from the same 'flesh'. The central question of genetic theory – the persistence of traits in sexual reproduction – was kept outside of the 1930 act by the restriction to asexually reproduced plants. As long as breeders managed to stabilize traits, however, they did not care whether they were acquired or inherited. After all, they were breeders, not geneticists.[8]

Indeed, the ontological status of genes remained shaky for some time to come. In the beginning, the central object of genetics was the chromosome. It became a separable object at the end of the 19th century, when cell biologists investigated the cell nucleus and noted that it was split into several fragments during cell division. These stainable fragments were called 'chromosomes' in 1888 by Heinrich Wilhelm Waldeyer. For the longest time, however, cytological genetics remained a descriptive science, which sought to understand the function of chromosomes and their role in the hereditary process (Müller–Wille and Rheinberger, 2012, pp 127–160). Experimental tinkering was possible (Curry, 2016, pp 77–129), but manipulation in a controlled and targeted fashion was out of reach. Plants were studied and described in terms of their chromosome numbers (Santesmases, 2013), experimental crosses (e.g., between wheat and rye) could be examined for possible translocations of chromosomes or their parts, but chromosomes could not be appropriated materially by picking them up and moving them in and out of organisms (see Hacking, 1983, p 23).

Much of the function of chromosomes remained unclear, but experiments gave plant geneticists some hints and evidence. Observing chromosome translocations under the microscope made it possible to link pathogen resistances in progeny of experimental crosses to a piece of the chromosome. This enabled cytologists to write fragmentary histories of trait inheritance. But for turning genes into fully-fledged scientific objects, more was needed. The identification of DNA as the substance which chromosomes were made up of in 1943 was a first step, but the universally accepted breakthrough of genetics was James Watson's and Francis Crick's (1953) revelation of the microstructure of DNA ten years later (Fox Keller, 2002, p 3). It was a breakthrough insofar as it signalled the end of the gene as an object inferred: in contrast to the chromosome, genes could only be deduced from the reconstruction of hereditary trees and statistical distributions of phenotypical traits, but not shown and photographed (Fox Keller, 2002, p 67). Among others, this ontological status allowed Barbara McClintock to postulate the existence of jumping genes (Fox Keller, 1983), but not to link them to a material object precisely located in space and time (Wieland, 2011, p 259). This was an issue because George Beadle and Edward Tatum (1941) had postulated a direct link between genes and enzymes, a connection that could potentially be extrapolated to phenotypic traits: for a specific characteristic of an organism, geneticists expected to find a cause in the form of a gene – only what form exactly the gene had was up for debate. Until Crick's and Watson's groundbreaking paper, genes thus enjoyed an existence comparable to those of atoms at the end of the 19th century (Stengers, 2011, pp 3–16): by and large, they were theoretical objects.

With the identification of genes with DNA, however, the former moved from what John Dupré (2004, p 326) calls 'Mendelian' to 'molecular' genes. Mendelian genes require an observable trait (like petal colour) whose distribution across individuals (such as pea plants and their offspring) is then compiled in a statistic. This statistic allows to inquire the paths this trait takes down the hereditary flow sheet (one fourth of a generation will inherit recessive colouration compared to three quarters with dominant colour alleles). One consequence was that Mendelian genes had to be inferred 'backwards' from the phenotype but remained statistical effects. With Watson's and Crick's paper, however, they turned into chemical compounds of a specific composition and structure. This should become crucial for patent law a few decades later: as Calvert and Joly (2011) point out, gene patents could claim an older tradition of patents on chemical compounds by referring to the chemical nature of the gene. As the gene had become a strand of DNA, it could now be expressed in a language that was no longer purely descriptive (like early plant patents which had to meticulously describe a phenotype in full detail) but which could reduce it to a simplified, essentialized formula that at the same time was more precise than phenotypical descriptions.

Mobile genes

Before genes were able to enter the patent office, however, something else was needed. Identifying life with genes with DNA with chemical compounds did not turn them into a potentially patentable subject matter yet. On a technological level, genes had first and foremost to be rendered useful before they could become a patentable object. Without any prospective technological – and therefore commercial – value, inventors would not go through the painstaking process of patent application, any good economist would object. But the leap from description and prediction of genes and their activities to manipulation designates more than an opening of potential application and, therefore, commercial opportunities. Just like genes gain in reality by being linked to DNA and being expressed in standardized, reductionist notation, they gain even more reality by being touched, cut, moved around and reattached (Hacking, 1983, pp 22–27). This has implications for their status as property objects. The possibility to control and manipulate meant that genes lent themselves to a new mode of appropriation in which the rights conferred by a patent could actually be exercised. The promise that hinged on this possibility was that with controlling the molecular gene, the Mendelian gene, by extension, could also be freely moved. It was the gap between description and manipulation that stood between genes as scientific objects and genes as technical objects; between genes as intellectual property in a Mertonian (1974) or Latourian (1984) sense and genes as patentable intellectual property with commercial potential.

For several decades, however, plant genes refused to become an object that would lend itself to this kind of technical appropriation. First of all, because the concept of the gene became increasingly complicated. While the relationship between molecular and Mendelian gene was originally assumed to be a one-to-one translation, it soon turned out that in a great number of instances, many-to-many was a much more adequate relation (Dupré, 2004, p 326). Neither were all genes just one continuous stretch of successive base pairs on a DNA string; instead they proved to be interspersed with stretches that coded for something else or even sequences which did not code for any RNA at all (Fox Keller, 2002, pp 59f). But if such 'details' *complicated* genetic theory, they did not necessarily *complexify* (Stengers 1997, pp 3–21) it: although they put some reservations, modifications and conditionalities on the idea of the molecular gene, they did not do away with the concept.[9]

Most importantly, other questions were of greater interest for the pioneers of genetic manipulation. One was whether organisms themselves could appropriate foreign DNA and incorporate it into their genome. In the early 1950, this had been demonstrated for three species of bacteria – *Bacillus subtilis*, *Hemophilus influenzae* and *Streptococcus pneumoniae* – but only for single-stranded, highly homologous DNA, that is, DNA which these bacteria already more or less possessed in similar form and which was therefore easily taken up by recombination with their own. These bacteria did so spontaneously and were generally seen as unspectacular exceptions to the widely accepted opinion that so-called horizontal gene transfer between species was impossible (Lurquin, 2001, pp 8–24).[10] In short, genes were not in a shape to be appropriated by foreign host genomes and host organisms were not prepared to accept and make use of new genetic properties.

Part of the obstacle was that little was known about the structure and function of the DNA sequences used at that time. Sequencing technology, nowadays widely used and cheap, was not available back in the 1960s and early 1970s. Knowledge about promoter and terminator sequences, which regulate transcription and replication of gene sequences, was fragmentary at best. From today's point of view, these early experiments lacked a sufficient concept of the gene, instead focusing on the incorporation and expression of undifferentiated foreign DNA strands in host genomes. They also assumed an easy and smooth transport of genes into host cells, whereas the following decade would be marked by the problem of finding suitable vehicles for taking up, moving and embedding genes in genomes (Lurquin, 2001, pp 54f). What was necessary for a successful translation, therefore, was a new material definition of the gene that would render it suitable for appropriation by a plant host genome – as well as cellular and molecular tool to pick it up.

The solution to the second challenge came in the form of a soil bacterium, *Agrobacterium tumefaciens*, which was known as the causative agent of crown gall disease, a form of tumour that a wide variety of plant species developed. Over the 1970s, a research race between the research group around Marc van Montagu and Jeff Schell at the University of Ghent, Belgium, and Mary-Dell Chilton's laboratory at Washington University in St. Louis, Missouri, ensued. It set off an avalanche of new findings about *A. tumefaciens*. The bacterium turned out to have the capacity to induce DNA from one of its plasmids (separate circular DNA strands within the bacterial cell) into plant cells which would then develop into tumour cells. Defined DNA strands from this so-called Ti plasmid would be incorporated in the cell nucleus and tumour cells exhibited hereditary physiological characteristics that could not be linked to the genetics of the host plant. In the course of this cascade of discoveries, *A. tumefaciens* emerged as 'a natural plant genetic engineer' (Lurquin, 2001, p 65).

Taking the bacterium as both their research object and their teacher, 1970s plant scientists learned what genes required to be successfully shuttled into a plant genome by *Agrobacterium*. It turned out that transposonal elements that were used to mark and track Ti plasmid DNA were incorporated alongside the bacterial sequences. Furthermore, one of the regions of the plasmid could be swapped for other sequences without disrupting the virulence of the transformation process (Lurquin, 2001, p 74; Wieland, 2011).[11] This made *A. tumefaciens* the ideal candidate for transporting genes into plant cells: there was some empty space in its freight hold, which researchers could use to load and ship genes of their own choice. One obvious candidate was a genetic resistance to antibiotics, widely used for research on genetically modified bacteria. In order to be recognized and transcribed within the plant's genome, however, these genes' sequences had to be 'translated' from a bacterial version into one that was compliant with the processing of transcriptional signals in plants (Lurquin, 2001, p 75).

Both *A. tumefaciens* and transgenes thus needed to be tamed and trained to enrol (Latour, 1993) them in the scientific programmes. From largely unknown, stubborn and reluctant entities, they had to be turned into familiar, well-behaved and compliant research tools (Wieland, 2011, pp 264f). This required enormous work, yet progress in the field marched fast. In the early 1980s, there was a close race between the Chilton group, the van Montagu–Schell laboratory and Monsanto's research division, which had only recently entered the challenge. In January 1983, at a legendary conference in Miami, all three teams presented their results, each indicating a successful translation of antibiotic resistance. This marked the first controlled transmission of select genes across the boundaries of taxonomic kingdoms, from bacterial to plant cells (Charles, 2001, pp 3–10).

An issue that remained, however, was the translation from plant cell to whole plant: as it was cell cultures, not fully-grown plants, that were used in these experiments, even a successful gene transfer would remain a dead-end street. Fortunately for Monsanto, they had enlisted Rob Horsch, a young plant scientist, who like no other could make plant cell cultures grow into whole plants again (Charles, 2001, pp 3–10). Monsanto was getting ahead of its public competitors, not only on the scientific, but also on the commercial stage: on 17 January 1983, the company filed for three patents on a method for transforming petunias with *A. tumefaciens* (US 5352605A; US 6174724B1; US 8334139B1). Having received note of the activities at Monsanto, their rivals Chilton, van Montagu and Schell had anticipated such a move. Both groups filed for patents themselves which culminated in three distinct patent applications (nonetheless claiming the same invention) at the US and European Patent Office in early 1983 (Charles, 2001, pp 20–22).

Life before the law

In the race to genetic transformation of plants we encounter a concurrent translation, different from the technoscientific ones we have encountered so far and which enabled plants to take the next step on their path towards patentable objects. Could all these developments as well have taken place a decade earlier? Charles Lurquin, who witnessed them as a contemporary at the forefront of plant science answers in the negative: it was only towards the end of the 1970s and early 1980s that cloning vectors and sequence manipulation techniques had become widely available, and only for well-known genes such as resistance to common antibiotics (Lurquin, 2001, p 79). Therefore, all earlier attempts to appropriate plant genes were doomed to fail, because they lacked an infrastructure or, rather, 'technological landscape' (Rheinberger, 1995, p 19) not just of epistemic but of technical objects that enabled certain movements and operations. Even if such an infrastructure had been available in 1973 already, however, it is very unlikely that the same achievements would have led to patents of the kind filed by Chilton, Monsanto, Schell and van Montagu.

Parallel to the scientific developments in the lab, a series of economic and legal translations had taken place which would result in a steppingstone for plants on their journey to property objects of a new kind. Setting a starting point for this parallel history is not trivial, as there are many predecessors which ultimately coalesced in the same event. One such predecessor has already been mentioned: the transsubstantiation of genes into DNA into molecules. This metamorphosis, however, did not immediately translate into DNA patents in 1953 or 1963. Back then, biologists never even considered patenting their results (Charles, 2001, pp 19f; Rabinow, 1996, pp 20–22). Science – especially plant science (Kloppenburg, 2004, pp 66–90; Fowler, 1994, p 132; Brandl, 2017, pp 175–198) – was not subject to the laws of the market and had little use for property beyond authorship in publications and citations (Merton, 1974, p 274f). Up to the 1970s, it was the state and its research institutes who funded and organized biological research, aligned some disciplines with its interests and left others to do as they please as long as their research did not interfere with politics (Dickson, 1984, pp 107–162). While European plant breeding was a notable exception in this regard (Brandl, 2017), US seed development and plant science was largely a public responsibility (Kloppenburg 2004, pp 50–90).

From the 1970s on, however, US plant breeding and genetics became a playground for private capital. Pharma multinationals (e.g. Hoffmann-LaRoche, Pfizer, Eli Lilly, Ciba-Geigy, Upjohn, ICI), petro- (e.g. ARCO, Shell) and agrochemical companies (e.g. Dow Chemical, Lubrizol, Monsanto) had started to invest large sums into plant breeding, molecular biology and their promises (Fowler, 1994, p 140; Kloppenburg, 2004, pp 211–214). The United States' new policy of using grain as an instrument of foreign policy towards socialist and developing countries raised hopes in the agricultural sector and led to a boom across the industry. Pharmaceutical companies entered genetics by expanding their biochemistry departments (Rabinow, 1996, p 27), agrochemical firms saw the potential of the agri-trade networks they used for pesticides and fertilizers for new seed products while the petrochemical industry was loaded with oil money. As Cary Fowler explains:

> Drug and petrochemical companies found that seed companies fitted well into their corporate structures – they could feel comfortable with the need for research, testing and such. Finally, the attention being focused on the need to increase food production drew constant to the seed industry. Any production increases would almost inevitably call for more or better seed from the seed industry. All of these factors, plus an attractive rate of return within the seed industry, encouraged takeovers. (Fowler, 1994, pp 132f)

The various corporate newcomers to plant science of the 1970s and 1980s had in common that they did not only have scientific experience with 'old' chemistry. They were also used to running it as a business, which they sought to extend to the new chemistry of DNA strands and sequences. On the one hand, molecular biology appeared as a continuation of their business models, but on the other it also represented a translation of these models into a new field.[12]

However, what all these companies required for pursuing their economic rationalities was first of all an instrument to translate technological breakthroughs into economic assets from which commercial products could be developed or value could be derived in other ways. Patents were the obvious choice for such an instrument: chemical companies were used to patenting their newly developed products and compounds; they had standing armies of patent lawyers and business models developed to turn patents into money. Unsurprisingly, early DNA patenting practices looked just like older practices in chemical patenting. They used the same language, the same ontology and even the same US Patent Office departments (Eisenberg, 2006, p 318; Calvert and Joly, 2011).

As previously seen, however, the commonly held opinion up until the 1970s was that living beings and utility patents did not go together. Even worse for commercial plant genetics, legislators had not only answered the question of patentability of plants in the negative back in 1930. They had also given an alternative answer in the form of plant patents (which were not quite what the chemical industry was looking for). Even the passing of much 'softer' PVP laws in 1970 had proven highly difficult for the plant breeding industry (Fowler, 1994, pp 148f). The promises of genetically manipulated plants were facing a wall. In 1980, however, a breach opened that would make a way through that wall possible, with only a slight detour. Two patents – one on an oil-decomposing bacterium that would become famous as General Electric's (GE) Chakrabarty patent and a much-forgotten one on a strain of antibiotics-producing bacteria by Upjohn – had been filed in 1972 and 1974, respectively, with the US Patent Office. The US Patent Office had rejected both, arguing that they claimed unpatentable subject matter (Kevles, 1994).

This line of argument was based on two grounds, echoing two long-standing reasonings from the political-economic and the moral-ontological tradition. The first made a point that bacteria were products of nature, a category for which there was sound legal precedent (Pottage and Sherman, 2010, pp 171f). The second argued that bacteria were living beings and therefore not patentable (for which there was no such precedent). This led to legal disputes that went on for years and finally reached the US Supreme Court in the late 1970s. Realizing that the fate of patents on organisms was tied to these cases, Upjohn made a strategic decision and dropped its lawsuit so that the Chakrabarty patent would become the test case for the established legal framework (Fowler, 1994, p 149; Kevles, 1994).

At this point, the fact that all the patenting issues plants had originally encountered in the early 20th century had been condensed into the category of 'nonpatentable subject matter' played out to the advantage rather than the disadvantage of the patent applicants. The reification of *practical* problems – disclosure, inventiveness, repeatability – into an *ontological* argument – that organisms were, by their nature, unpatentable – turned all these issues into a single one. Instead of fighting on a number of different battlegrounds, GE only needed to defend its patent in one arena by delivering a compelling argument why its bacterium was eligible for patent protection. The same would apply to the court's majority opinion if it were to rule in favour of GE (as they eventually did) and explain their reasoning. In its alleged solidity, the question of patentable subject matter was highly misleading, because it black-boxed technical questions and reduced them to a seemingly more fundamental matter. While it resonated well with public sentiments that organisms could and should never be patented (Parthasarathy, 2017, pp 55–59), the fact that it was an ontological rather than a practical or technical argument did not make it more but less irrefutable.

This is apparent in the opinion of the majority of the judges who ruled in favour of Chakrabarty and GE:

> Here ... the patentee has produced a new bacterium with markedly different characteristics from any found in nature, and one having the potential for significant utility. His discovery is not nature's handiwork, but his own; accordingly it is patentable subject matter under § 101. ... Two contrary arguments are advanced, neither of which we find persuasive. ... The petitioner's first argument rests on the enactment of the 1930 Plant Patent Act ... and the 1970 Plant Variety Protection Act ... In the petitioner's view, the passage of these Acts evidences congressional understanding that the terms 'manufacture' or 'composition of matter' do not include living things; if they did, the petitioner argues, neither Act would have been necessary. (Diamond v. Chakrabarty, pp 310f)

The court majority clearly dismisses the claim that genetically modified bacteria are products of nature here: there are, they argue, no naturally occurring bacteria able to decompose oil; hence – and here is the first leap – it must be a patentable invention, as it must have been Chakrabarty's doing which produced the bacterium in its form.

Furthermore, as the bacterium was a material object, it was covered by Article 101 of the US Patent Act, which lists 'manufactures' and 'compositions of matter' as patentable. That organisms, by their nature, were outside the scope of patent law was negated by the majority of judges with reference to intention of US Congress in 1952 that 'anything under the sun made by man' should in principle be eligible for patent protection, a definitive departure from patent law as a *sui generis* IP right for mechanical inventions (Pottage and Sherman, 2010), or at least what was left of that character. The existence of the 1930 Plant Patent Act did not contradict this in their eyes; it was an expression of practical problems of disclosure which had been overcome (although they did not state how), not of essential exemption (447 US 303 (1980), pp 2206f, 2210). In contrast, legal questions arising within the realm of the living such as who or what actually reproduces and practices an invention if a patented bacterium splits (Braun and Gill, 2018) were not addressed.

Undoubtedly, Chakrabarty's *Pseudomonas* bacteria were not 'made' by him in the sense of being assembled from scratch; 'transformation' may be a more accurate term for the procedure they underwent.[13] But this was not what the court's majority sought to address: things (including organisms) could either be 'in nature', in which case they belonged to all and none, or they could be absent from nature, in which case they were perfect candidates for becoming someone's property. Whether that happened in an analogous way to the assembly of machines or the synthesis of chemical compounds was of secondary importance in the *Chakrabarty* case. What was at stake was the ontological status of organisms as inventions that could be appropriated via patents, which the Supreme Court answered in the affirmative.

It is difficult to overestimate the significance of the *Chakrabarty* ruling for biopatents (Eisenberg, 2006), but it is equally difficult to assess its specific role within the multitude of paradigm shifts of the 1970s and 1980s. Not that it came up with the idea that organisms could be patentable: while such an idea was foreign, unheard of or scandalous for parts of the public (Fowler, 1994, p 149; Jasanoff, 2007: 48–50; Parthasarathy, 2017, pp 55–59), the case was a result of a growing number of patent applications on new biotechnological objects. These patent applications testify to the fact that, unlike in 1950s and 1960s 'Big Science', patents on organisms were not alien to researchers, at least in the private industry. Nonetheless, *Diamond v. Chakrabarty* represents a watershed, as it signalled a new era in which such 'designs on nature' (Jasanoff, 2007) would be widely sanctioned. Although it was submitted before the ruling of the Supreme Court (in 1978), it was only after the decision in *Diamond v. Chakrabarty* (in 1982) that the first patent on a molecular gene was granted by the US Patent Office (Calvert and Joly, 2011, p 5). Until the beginning of the Human Genome Project, such DNA patents were not met with considerable opposition by the public (Eisenberg, 2006, p 318).

It was not only molecular gene patents, however, which saw a surge after the *Chakrabarty* ruling: utility (i.e., classic) patents on plants (as early as 1985) and animals (from 1987 on) were subsequently granted by the US Patent Office, which argued that (except for humans) an organism was an organism (Fowler, 1994, pp 150f; Parthasarathy, 2017). Had the issue in *Chakrabarty* not been framed as one of eligible subject matter, but more practically as one of the biological problems of appropriation from the earlier debates over plant patents, the Supreme Court may still have ruled in favour of GE – but it would have been more difficult to construe the decision as a green light for patenting organisms at large.

The space for patentability in the US was now wide open; the realm of the living was legally annexed to the empire of patents. Despite the leaps in the reasoning of patent advocates and the break (rather than shift) in paradigms that *Diamond v. Chakrabarty* may represent from today's point of view, it is important not to understand the ruling as a scandal in which courts and Big Industry conspired to appropriate life on earth through patents. I am sceptical, for example, whether it can be read as a test case for democracy in science, in which the US Supreme Court at the same time drastically slashed and overstepped its own political powers in the matter (see Parthasarathy, 2017, pp 52–62). Nor should the contingency of the event (apparent in the close five to four majority of judges) encourage us to stop short at speculating what today's world would look like had the scales been tipped in the opposite direction.

If we can draw a lesson from *Diamond v. Chakrabarty*, then it is one about legal concepts and their malleability, but also their dependency on the outside world. Ten or 20 years earlier, the case would have taken place in an economic (and therefore political) vacuum (Jasanoff, 2007, pp 48f) – and, even more importantly, in a material one. As we have seen, genetic engineering remained a pipe dream for much of the 20th century, and technologies of disclosing and referencing microorganisms were developed alongside them (Pottage and Sherman, 2010, pp 200–206). Genes required a molecular ontology before they could be treated analogous to molecules (Eisenberg, 2006; Calvert and Joly, 2011) and an infrastructure of molecular and microbiological tools before they could be grabbed and lifted like Locke's acorns.

Plants had to wait for the vanguard of microorganisms to enter the domain of eligible subject matter. Legal translations, however, were just as crucial: without the precedent of chemical compound patents (Dutfield and Suthersanen, 2005), the gulf between genetic objects and patents' original domain, mechanical engineering, may very well have been too broad to bridge (Braun and Gill, 2018). The legal mode of operation by analogy requires established precedents to construe similarities as well as differences; it is dependent on footholds provided by statutory texts and statements of legislators, but at the same time confined by the condition that there must be no contradictions between the constituents of the law. It has to determine whether there are openings that allow for accommodating a novel element – such as treating microorganisms as inventions – without entering into disagreement with any other element (Latour, 2013, pp 64–69).

Obviously, every time something is translated, certain aspects are also lost in translation (Braun and Gill, 2018). Microorganisms are not manufactures in the way that assembly-line-produced calculators or electric toothbrushes are manufactures. They may well be 'compositions of matter', but they cannot be reduced to them. At the same time, the concepts we use for translating, such as patents, are transformed as well: whereas in the late 19th century patents were largely limited to synthetic chemicals, technical processes, or machines (Pottage and Sherman, 2010, pp 153–182; 2011), they also extended to modified organisms after *Chakrabarty*. The US Supreme Court's ruling represented a step in a long metamorphosis of patents in which central criteria such as eligible subject matter, non-obviousness or disclosure have been side-lined in favour of utility, as Rebecca Eisenberg has noted:

> Since *Diamond v. Chakrabarty*, the Federal Circuit has gradually abdicated its authority to police these boundaries in favour of an approach that collapses the traditional restrictions on patent eligibility into the single requirement that the invention be 'useful'. The PTO has ultimately acquiesced in this liberal approach to the threshold requirement of patent eligibility, diminishing the likelihood that the issue will come before the courts for review. (Eisenberg, 2006, p 318)

Patents, as we will see in the following, have gradually become an instrument that may appropriate anything as long as it is useful, whereas the obligations traditionally tied to them as part of the 'patent bargain' (Biagioli, 2019) have become less and less important after *Chakrabarty*. The fusion of genes and organisms in cells has at the same time proven to be a versatile and fruitful articulation for further ontological mutations. But before we can explore such subsequent legal and material shifts, we have to follow another translation; this time across the Atlantic.

Plant patents are coming to Europe

As seen in Chapter 3, Europe had taken a different path after World War II, opting for a *sui generis* IP regime for plant innovations rather than expanding patent law to plant breeding. For several decades, this system worked well, or at least to the satisfaction of the stakeholders that had been taken into account by the creators of PVP law (Brandl, 2017, pp 129–130). With the advent of biotechnology, however, Europe, too, was faced by new plants and the question of how those should be appropriated. Although to a lesser extent than the US, the European seed sector had witnessed an influx of money from the chemical and pharmaceutical industry (Parthasarathy,

2017, p 51); domestic players like Hoechst had entered the global race for genetically engineered plants (Charles, 2001, pp 70–73) and pressed for an adequate legal framework.

The legal landscape of Europe, however, was much more fragmented and complicated than the US one. It had taken the countries of Europe 24 years (from 1949 to 1973) to finally pass a European Patent Convention (EPC) that would allow patent holders to file one application for all its signatory states. This convention and its EPO were not organs of the European Economic Community, although like the latter, they sought to harmonize market access. There was some overlap between the signatory countries of the EPC and the member states of the EEC, but not all the EPC signatories (for example Switzerland) were or are members of the EEC's single market (Parthasarathy, 2017, pp 38–49). While today, all member states of the European Union (EU) are also part of the EPC, this was not always the case throughout the various waves of EU enlargement.

The EPO, with its offices in Munich, The Hague, Berlin, Vienna and Brussels, is not subordinated to the political institutions of the EU such as the European Parliament or the European Commission. Instead, it acts as an international organization based on the EPC. Since its jurisdiction coincides to a large part with the EU's, however, both sides seek to harmonize their decisions and regulations. There are several bodies of laws and regulation guiding and governing the EPOs decisions and rulings: aside from the EPC, the EPO abides by the so-called implementing regulations which, as a form of internal codex, reflect certain laws and previous decisions; it constantly seeks to harmonize its position with that of important EU institutions; and, important for the context discussed here, its approach to biotechnology patents follows an EU directive (Bently et al, 2018, pp 423f). However, its decisions are not subject to the European Court of Justice and cannot be appealed other than at the EPO itself. Together with a treatment of the EPO as an 'international organization' that comes close to the legal immunity of embassies and diplomats, this fact has been used by critics to call the legal status of the EPO within the EU into question (Broß, 2014; Then and Tippe, 2016, pp 9–11). As will we will see, the exact relationship between the EU and the EPO is an ambiguous one.[14]

The original EPC allowed patents on life forms (Parthasarathy, 2017, p 62) as long as they did not refer to 'essentially biological processes', animal races or plant varieties (EPC 1973, Art 53(a)). The reason for this exclusion, at least of plants, lay in the fact that the UPOV convention had existed for over a decade by then, was widely accepted, and still contained a mutual exclusivity clause for PVP and patent protection (Bently et al, 2022, pp 514–519). Although the EPC was already in force when biotechnology set out to change plant breeding, it did not become the major arena in which patentability of plants in Europe was defined. It was the European

Commission which, in 1988, took it upon itself to provide clear and coherent rules for the patenting of biotechnology after industry had called for an explicit framework. Aside from parliamentary public concerns over life-form patents, which were more vocal Europe in the 1980s than in the early US in the 1970s (Partasarathy, 2017, pp 62–79), the UPOV system was one of the main obstacles the commission encountered.

Here was a legal problem quite different from the one that US courts had to deal with regards to life-form patents: eager to catch up with the American industry and market, the commission was intent on granting genetically modified organisms (GMOs) the same level of patent protection they enjoyed in the US. PVP laws, however, explicitly allowed competitors to use protected varieties for research, breeding and developing new varieties (reassured in the commission directive 2100/94, which harmonized PVP in Europe, in 1994). If, for example, a commercial maize variety contained a transgene for insect resistance, patent law held that that resistance gene could not be used for cross-breeding without consent of the patent holder whereas PVP stated that the variety (including the transgene) could be freely crossed without permission or compensation. Since any seed to be commercialized in Europe had to be of a proved (and therefore protected) variety, any transgene to reach the market would actually have to be included in a protected variety.

Genetically modified plants summoned a complicated collision of alienation (seed market and cultivation admission), inclusion (breeder's exemption) and exclusion (patent law). The authors of the Biotechnology Directive found themselves in a dilemma: either patent law's extensive privileges would trump PVP's breeders' exemption, or vice versa – there was no way of squaring the circle. The commission could not simply do away with PVP, on which a whole industry relied; nor could it implement the desired changes into its own PVP directive, since genes were too different an object from plant varieties. Undoubtedly, the reasoning of the directive's authors centred on genetically modified plants, which had their heyday in the 1990s. Economic value was concentrated in expensive transgenic traits such as Bt toxins or resistance to glyphosate, which had been developed, patented and commercialized in the US and were expected to hit the EU market soon (Charles, 2001, pp 149–170). In 1998, after ten years of deliberation and consultation, the commission found a solution: transgenes in and transgenic plants as such could be patented, but plant varieties remained exempt from patent protection. The most crucial detail of the directive, however, lay in one of the considerations that parliament and commission included as a sort of preamble: 'Whereas this Directive is without prejudice to the exclusion of plant and animal varieties from patentability; whereas on the other hand inventions which concern plants or animals are patentable provided that the application of the invention is not technically confined to a single plant or animal variety' (Directive 98/44/EC, art 29).

This meant that a patent application claiming a trait confined to a single variety was not admissible – if, however, the application claimed applicability in two varieties or plant species, it was compliant with the Biotechnology Directive. The directive thereby took the opposite stance to a popular legal narrative (Bently et al, 2022, pp 514–517) that depicts PVP as an IP right protecting the variety 'as a whole' (that is, as a sum of traits) and traits or genes as 'parts' (that is, single elements that could be included in a variety). This ontology was upside down: in the directive's understanding, patents on genes could contain varieties, as the utilization of traits extended beyond variety and species boundaries.[15] No matter what the exact hierarchical nature between traits and varieties, patents and PVP, the Biotechnology Directive, by spelling out their exact relationship as not strictly mutually exclusive, rendered them legally compatible.

Despite the quick implementation of the directive in several national patent laws, however, the Biotechnology Directive did not lead to a boom in agricultural biotechnology. What stopped transgenic plants was not a lack of property rights, but the inability to turn one of them – the right to alienate – into a fully-fledged economic script of commercialization. While legislators were busy understanding genetically engineered plants as an object of technicality, invention and economic value, and firms like Monsanto had praised them as the solution to hunger and environmental problems, an anti-GMO movement emerged across Europe. For this broad movement, genetically engineered plants did not only embody all the ills of modern capitalism, but first and foremost posed an environmental and health risk. Plant science companies, molecular biologists and other proponents of green biotechnology soon realized that without cooperation and consent of consumers, farmers and supermarket chains, their patents were more or less worthless (Wynne, 2001; Jasanoff, 2007; Bonneuil and Thomas, 2009, pp 406–408). While anti-GMO activists had largely lost the patent battle over whether and how transgenic plants could be appropriated (Parthasarathy, 2017, pp 51–79), they defeated the alienation of GM food in the fields of scientific expertise (calling into question its safety) as well as consumer and environmental protection law (Gill, 2003, pp 239–244; Charles, 2001, pp 205–229, 236–262; Jasanoff, 2007, pp 117–145).

Patents without genes

The story of European patents on plants could have ended here. With GM food a dead horse in Europe, researchers warned that research might be pulled to more GM-friendly places, and indeed plant science firms stopped their biotech research and product lines in Europe. To this day, European consumers remain widely opposed to GM foods (Bonny, 2003; 2014; Hicks, 2017) and tend to associate plant patents with green biotechnology, dismissing both in one go. However, although it might well have been otherwise, GM plants and the EU Biotechnology Directive did not become a dead end for patents on plants. Rather, it turned out to be a catalyst for an even further expansion of the realm of patentability. The reason lay in the legal architecture of the various bodies of law that governed patentability in Europe as well as the wording of the Biotechnology Directive – which came as a surprise for traditional plant breeders, who had thought that the Biotechnology Directive would only have implications for biotech plants.[16]

What exactly had happened? Although careful to keep the domain of traditional plant breeding – varieties and PVP law – out of the scope of patentability, the Biotechnology Directive's effort led to the opposite result. The clear delineation and definition of plant varieties had the effect that the scope of patent law was in turn understood to comprise everything not covered by PVP. The arena for defining PVP and expanding plant patents in 1990s' Europe was the EPO with its two boards of appeal. In *Ciba-Geigy's Application* (decided in 1984 and frequently cited in later cases), its Technical Board of Appeals (the EPO's first instance for appeals) ruled that plant material with certain characteristics obtained by chemical treatment did not qualify as plant variety because, among other things, such characteristics were not passed on over generations – hence, the plant material was patentable. In *Novartis/Transgenic plant* (1998), the Enlarged Board held that if patented transgenes were included in a protected plant variety, this did not contradict existing PVP or patent law (Metzger and Zech, 2016, p 634; Bently et al, 2022, pp 520f). In restricting the category of plant varieties to its minimum definition – that of standardized seed fulfilling the DUS criteria – the EPO created new legal space to be filled with patents.

This opening mattered less for biotechnology like marker-assisted breeding (MAS) and genetic engineering than it did for 'low-tech' breeding. Some companies realized that if only varieties were outside the scope of patent protection, they could try with breeding methods. In itself, this approach was not too far-fetched: hybrid breeding systems, for example, had been subject to patent protection for a long time and were not strictly considered 'essentially biological processes' in the EPC's sense. Similarly, the increased use of molecular markers in plant breeding (Reece and Haribabu, 2007) made it possible to patent molecular compounds that would greatly speed up the selection process. Simple cross-breeding and selecting, however, was usually regarded as the embodiment of an essentially biological process (Braun, 2023), acknowledged already by Charles Darwin in his *Origin of Species* (Darwin, 2008, pp 63–68).

Echoing the epistemological problems with plant patenting, biological processes were long considered inherently non-reproducible in European patent law. This was the reason, for example, why the German Federal Court ultimately dismissed a patent application for a pigeon breed in the so-called

Red Dove case in 1969 (Metzger and Zech, 2016, p 631). If plant patents were sporadically issued before the introduction of PVP, it was because the biological material itself (not the breeding method described in the patent specification) was considered enabling, that is, giving third parties the necessary means to reproduce the invention.[17] Although the Budapest Treaty of 1980 had created an international framework for describing and depositing biological material for patent purposes (Pottage and Sherman, 2010, p 190), referring to biological processes by either textual or material means in an enabling way long seemed impossible.

This did not deter the State of Israel (as representative of its public research programs) and the UK company Plant Bioscience from filing patents on breeding methods for selecting for self-drying in tomato (State of Israel; the *Tomatoes* patent; EP 1211926) and increased glucosinolate content in broccoli (Plant Bioscience; the *Broccoli* patent; EP1069819) in 2001 and 1999, respectively. The description of the breeding process claimed was rather simple and generic; the claims were wide in turn: the claims to the process each amounted to cross-breeding between closely related species of *Lycopersicon* (tomatoes) and *Brassica* (oilseeds), respectively. In this setup, one of those species was of commercial and culinary importance, whereas the other contained desirable characteristics of agronomic or nutritional importance. This is and was a standard procedure in plant breeding, especially for obtaining resistances against pests. Most notably, neither of the patents included any hint of genetic markers, gene sequences or even biological material deposited as reference at a gene or seed bank. Nevertheless, patent examiners at the EPO deemed both processes inventions and hence granted the patents in 2002 (*Broccoli*) and 2003 (*Tomatoes*). Immediately, protest ensued and two somewhat unexpected actors used the possibility given by the EPO to file for opposition: the Swiss plant science multinational Syngenta (joined by the French breeding company Limagrain) objected to the *Broccoli* patent, while the Dutch food giant Unilever opposed the *Tomatoes* patent (Braun, 2023).

Why did such mundane patents cause so much resistance, of all things? The reasoning of some the actors involved will be discussed in Chapter 5, but here it is probably necessary to point out that the patents had major implications for plant breeding: not only did they cover a method many experts regarded as perfectly obvious and explicitly exempt in the Biotechnology Directive, they also claimed the products of this process. The plants obtained by such a breeding method would therefore constitute patented objects. This meant that for the first time, conventionally bread plants had entered the realm of patentability in the EU, which up to that date was reserved for biotech plants. While biotech traits were few and effectively 'extinct' in the EU, so-called native traits were the daily bread of the breeding industry, proliferating widely across company germplasm pools by way of the breeder's exemption. But beyond such far-reaching implications for business, the patents also signalled that the exemption of plant breeding and its subjection to a distinct IP regime was eroding away. If future plants were to include not just one or two transgenes, but a thicket of patent-protected native traits, PVP would become increasingly meaningless as a result.

It took the EPO a several years to come to a first conclusion in the two cases. The Technical Board of Appeal referred the cases to the Enlarged Board of Appeal in 2007 and 2008, which in late 2010 decided that the methods described in the patents could not be patented as they constituted essentially biologically processes (G 2/07; G 2/08). The ball was back in the Technical Board's field. In the face of the Enlarged Board's decision, it signalled to the patent holders that the methods claims could not be upheld – but that they could retain their material claims, provided they redrafted their patents accordingly. Again, it referred to the Enlarged Board of Appeals in 2012, this time asking whether they regarded the tomato and broccoli material obtained by essentially biological processes to be outside the scope of Article 53b of the EPC. In March 2015, the Enlarged Board finally ruled that the exclusion of biological processes did not extend to their product and that both patents in their amended form could be granted (Braun, 2023).

The legal status of plants in 2015 was thus somewhat paradox: in the US and many other countries, they had been perfectly patentable at least since the *Chakrabarty* ruling. Conversely, in a confusing European geography constituted by several overlapping but not perfectly congruent jurisdictions – EU PVP law, the EPC, the EPO's implementing regulations, national patent laws, the Biotechnology Directive – plants as property objects moved back and forth. There was some agreement over what was firmly inside the boundaries of PVP and patent law, but the space in between remained fuzzy. Repeated attempts by patent proponents sought to get as close to the letters of the exemptions in Article 53 EPC to fill that legal space with patentable object matter. While they were not always successful, were largely sanctioned by the rulings of the EPO – an attitude the legal scholar Axel Metzger likened to having opened 'the floodgate' (Metzger, 2016, p 516).

Conclusion

Whether the decision in *Broccoli/Tomatoes II* was justified or a major legal error, it constituted a passage point in the biography of plants as property objects. In a sense, plants have gone full circle: originally, plants were the products of more or less mundane breeding practices, often just as unspectacular themselves. Two peculiarities prevented them from being turned into patentable objects: the impossibility to translate them into explicit knowledge – be it as some form of recipe or doctrine, be it a reductionist representation that captured the essence of the innovation – and the unorthodox distribution of innovative agency between 'inventor' and 'invented'. Even though Europe and the US pursued different paths after they had come to this conclusion – one making PVP the cornerstone of its breeding industry, the other simulating patents for single traits to the extent possible – their abandonment of classic patent law for plants demonstrates that plants were not only legally but also materially and technologically outside the ontology of the patent.

A series of events, however, would transform plants in ways that made them appropriable through patents. One major development was the molecularization of life, opening the possibility of defining plants in molecular and informational terms. It took the techniques of genetic manipulation, however, to turn plants and their genomes into objects of technology. The changing biology of plants thus became a driver, rather than just an object (Landecker, 2016), of propertization. None of the applications of genetic engineering had been foreseen by the Plant Patent Act, PVP laws or patent law. And yet it was the latter that was deemed most suitable by plant science companies and courts to deal with these novel plants. Nevertheless, the nature of plants as patent-eligible subject matter still had to be established. It was the decision in *Chakrabarty* that finally circumvented the reservations against 'real' plant patents brought up 50 years earlier: not so much by refuting them but by dismissing the unique nature of life that the USPO and patent critics had brought up as their main argument. A curious instance of loss in translation can be encountered here: while genetic engineering and molecular biology were necessary to build up a case like *Chakrabarty*, they were no longer needed to establish patentability in its aftermath. They served as a catalyst for the convergence of patents and conventionally bred plants, which became redundant after the latter had turned into patentable objects. Plant innovations of all kinds could now be patented with utility patents instead or on top of plant patents (or PVP titles), an option US seed companies use extensively to this day (Janis, 2017).

European legislators sought to follow the US model with regard to biotechnology while keeping conventionally bred plants outside of the scope of patent law. Ironically, however, the victory march of genetically engineered plants was halted by popular resistance, whereas the EU Biotechnology Directive opened a backdoor for patents on non-transgenic ones. The uneasy relationship between varieties and traits and the priority ultimately given to patent law as the more universal intellectual property regime eventually led to the *Broccoli/Tomatoes II* case at the EPO. With the EPO's eventual decision to grant both patents, plants had passed from essentially biological to technical and back again in Europe, too: they had acquired a molecular, transgenic identity which built up to the drafting and passing of the Biotechnology Directive, which paved their way into patent offices – once there, they could afford to shed this identity as quickly as they had acquired it, returning to their 'essentially biological' identity. This way, plants as patentable objects did not only survive the tumultuous era of the 'GMO wars' (Stone, 2002). Subsequently in its aftermath, patent law could also readily claim all essentially biological products as its legitimate domain.

Some episodes of this history appear predictable in hindsight: parallel to plants, patentability has expanded to include 'inventions' such as business methods, genes and algorithms – at least in some jurisdictions (Eisenberg, 2006; Pottage, 2006). One way of making sense of the *Broccoli/Tomatoes II* story would thus be to see it as part of a 'long march' of patents. Conventionally bred plants would then only be the most recent stage in this history, while propertization would continue to expand with patentability. Such a view would disembody the process at the same time: it would remain unclear what propels the expansion of the patent kingdom. Perhaps we could attribute the victory march of patents (and other intellectual property regimes) to the forces of capitalism in general, to business interests in particular or simply to propertization as one of the defining trends of modernity (Maurer and Schwab, 2006).

Siegrist's (2005) second definition of propertization – that of a self-driven, rather than instrumental process – could be read in this vein. In this understanding, propertization continues to shape the world as an unmoved mover. I have interpreted it differently here: as a process that does not rely on anyone's plan but that still cannot move by itself. To borrow from Gibson-Graham (2006, pp 9f), propertization may be a monster, but it is not a fabled beast: it still needs to feed and sleep. In the history sketched in this chapter, plants had to come a long way before they could legally be grasped as inventions; a journey that took even longer in Europe than in the US and went by different ways. Plants had to be translated various times over to get where they are now, relying on scientific breakthroughs, economic change and technological advances. But patents, too, had to change: only by forgetting their origins in mechanical engineering they claim new territory (Pottage and Sherman, 2010). As a result, a patent means something different today than in the early days of modern plant breeding. Patents have increasingly abandoned 'invention' for 'innovation', side-lining the formerly central figure of the inventor and her creative act (Machlup, 1958, p 77; Kang, 2015, p 30; Godin, 2017).

Not all of these translation events can be read as anticipating the propertization that would follow them: we might just as well understand the history of patentable plants as a long series of accidents. There is no 'great plan' for patents, harboured by Big Industry patent lawyers or the logic of capitalism. If anything, there is the coalescence of particular interests, the appearance of new objects and techniques and suitable political climates that allows propertization to proceed. Power is no unimportant part of this history of property but neither predicts nor explains it (Bonneuil and Thomas, 2009, pp 10–25). There are, of course, certain logics or imperatives to things themselves: patents suggest and require invention at the same time; genes offer a link between traits and chemical compounds; competition demands that Europe follows a North American model. Yet these are all far from unified or straightforward, as we will see in the next chapter. We can and should still ask 'cui bono?', but we should not confuse the benefiters with the protagonists of this history (Star, 1995).

Tracing the trajectory of propertization from now back to the past allows us to understand that a world without patentable plants – or copyright-protected books, trademarked fictional characters or privately owned land, for that matter – is thus conceivable. As self-evident, relieving or convenient this statement may be for the critics and opponents of patents on plants, it also goes the other way around: there is nothing in the nature of plants or patents that would eventually preclude them from coalescing with patents. Plants have challenged and continue to challenge the concept of the patent in many ways through their biology and social ecology. Yet that does not mean that patenting would be against the nature of plants or life itself. What the history of propertization in plant breeding teaches us is that neither the possibility nor the impossibility of patents – and therefore property – should be taken for granted.

5

The Values of Patents

What is the value of property? When we speak with Marx (1991, pp 886f) about property as enclosure or with Proudhon (1994, p 13) of property as theft, the underlying assumption is that property is valuable. The scandal of property, here, lies in the redistribution of its value; the fact that it is given to someone and taken from someone else – usually without compensation. Simple as it may seem, the argument is powerful. It challenges the established economic arithmetic of property by politicizing them: someone's profit is always someone else's loss. Far beyond its original context of 19th-century political economy, it has been successfully mobilized, from colonialism (Bhandar, 2018) to academia (Bollier, 2002), from music (Seeger, 2004) to medicine (Dickenson, 2017), from housing (Blomley, 2008) to seed (Kloppenburg, 2004). Drawing into question the idea that property simply produces wealth, the juxtaposition of winners and losers, of included and excluded allows us to rethink and criticize the role of various forms of property in economic and public life (Loick, 2023).

Useful as it may be elsewhere, however, the application of this critique to plant patents is not straightforward. The problem begins with an apparent lack of value of the *Tomatoes* and *Broccoli* patents, as well as similar applications. As seen in Chapter 4, both plants were strikingly low-tech, almost trivial in nature. What prompted the applicants to seek property protection for something that cost relatively little? Although they were ultimately granted by the EPO, the commercial life of the *Broccoli* and *Tomatoes* patents was short and underwhelming (Braun, 2023, p 59). At the same time, it is not clear who these plants should belong to, if not the patent applicants. What is being robbed here, then, and from whom? Why would anyone go to court over these patents, and why was it not the usual suspects – farmers, activists or small breeders?

The political-economic critique of property is underlaid by a strong notion of value. Not only does property have value, this value can also be distributed and measured. Moreover, it means something to both those who acquire it and those it is taken from. On an individual level, the distribution of property thus follows a zero-sum arithmetic: it is either mine or yours, or ours, but I can only increase my share at your expense, and vice versa. On a higher level, however, a different mathematics comes into play: here we can make claims about the unequal distribution of wealth (Piketty, 2014) or the tragedy of the anti-commons, in which we all become poorer (Heller and Eisenberg, 1998; Lametti, 2013; Aboukrat, 2021). In both cases, the weight of this critique derives from the weight of value – value that resides in the property object and can be transferred alongside it without changing.

We could also think of value differently. What if it did not lie inside the (literal or metaphorical) fences of the enclosure but outside of it? One alternative location of value could be the mind of the owner – it is she who projects value onto her property, rather than deriving it from inside her possessions. For example, Pierre Bourdieu's (1984) theory of appropriation can be read along these lines: here the educated middle class sees different values in a painting than its wealthy owner, who has lent it to a museum. For this reason, both can appropriate the same work of art and give value to it, albeit in very different ways. At the same time, popular and educated values diverge when it comes to art and culture – it takes more time and effort to properly value jazz than pop. Value, here, is subjective in the sense that it resides or at least extends from the appropriating subject. Property means something to its various owners, but it does not mean the same to them. If there is an arithmetic to it, it is that value can only be counted in the plural – values – and not be put in the same equation: ultimately, the market value of a piece of art is not commensurate with its cultural value (Karpik, 2010). But are we to believe that this is how companies think about intellectual property, investments and revenues? Do people come to appreciate the aesthetics of patents that belong to someone else? And what critique of property could we even build on this concept of value?

The two theories of value mirror two competing models of property. The political-economic one resembles that of property as an object, which can be measured and transferred. The second one, which we could perhaps call cultural-economic (du Gay and Pryke, 2002; Pratt, 2009), thinks of the value of property as something that ties to subjects rather than objects, or at least detaches more readily from the latter than from the former. It bears some resemblance to the property-as-relation model: urging us to look for value in people and their social position, rather than in the things they own in various ways. Either, however, will mislead us if we seek to understand the problem with native trait patents along its lines. The value of patented plants is neither strictly 'inside' the plants or the patents, nor does it reside 'outside', in malleable valuations that take place in the minds of their owners. Just like I have criticized both ideas of property earlier in this book, I want to problematize the two contrary notions of value in this chapter. What is needed instead of a substantivist or a constructivist notion of property value, as Fabian Muniesa (2011) has argued, is a pragmatist approach.

Native trait patents like *Broccoli* or *Tomatoes* offer an intriguing test case for this. While they seem to unite a range of very different stakeholders in their opposition, the respective reasons and motives for opposing (or endorsing) them diverge. That some of these stakeholders do not have patents does not mean that there is no value in these patents for them; while some who hold a vast patent portfolio see it as a liability rather than an asset. As we will see, it is not quite clear what is 'in' a patent (Bowker, 1992; de Laet, 2000), but it is not any easier to say what is outside of it (Biagioli, 2019). The controversy over native trait patents is both political and cultural: the various parties to the controversy have diverging, sometimes opposite ideas about the value of such patents. And yet, these values are made commensurate time and again by patents themselves.

Rewarding innovation

According to mainstream economic theory, patents are the prize that is meant to make people enter the innovation race (Machlup, 1958; Callon, 1994; Landes and Posner 2003, pp 294–333). 'Useful ideas, like crops, are generally the product of hard work, which must be rewarded in order to be encouraged', Posner (1986, p 54) writes. In this kind of economic justification, the prospect of getting a patent serves as an incentive for people to invest into R&D. Patents are needed because there is a discrepancy between the costs of developing a product and the price that people are willing to pay for it. While this is true for a lot of goods for which there is no market, the market does not *have* to fail here, the justification goes: the divergence between costs and market prices can be overcome by a combination of discounting and artificial, temporally limited monopolies. Successful inventors will be able to sell innovative and useful products exclusively for the duration of a patent, thereby evading competition and receiving higher profits which will help them recoup their initial investments. Without patent protection, however, competitors would undercut prices and rob innovators of their income: unlike the latter, they do not have to bear the costs of the expensive and risky initial appropriation of the idea by laborious experimenting and subsequent development (Machlup, 1958, pp 77f; Callon, 1994).

For companies, the potential value of such patents thus lies in the possibility to recoup expenses that would otherwise be lost to the market. Hubert Kempf, for example, could protect the fungal resistance traits he painstakingly accumulated in his nurseries and varieties from competitors: although it would not make breeding his wheat any cheaper for himself, a patent would keep others from selling the same resistances at a lower price. Once the patent expires, however, this 'legal fence' would be pulled down and the traits would become freely available to everyone else. These are the social benefits (cheap access to technology after the patent) meant to outweigh the social costs (higher prices before expiration). In the absence of patent, the argument goes, plant breeders would sooner or later stop investing in trait development (Landes and Posner, 2003, pp 312f) since their competitors would have free access to expensive traits.

In Michael Kock's words, such an arrangement would amount to 'killing the hen to get to the egg' (Kock and Gould, 2014, p 98). Until 2016, Kock was Head of IP for Syngenta, the Swiss agrochemical and biotech giant. He argues that patent can promote trait development where PVP cannot. "It's about preventing market failure", Kock says, "about saying, how do I create sufficient incentives; especially in a field where seed reuse and easy propagation would otherwise lead to a rapid loss of value?"[1] Kock sees an under-investment in trait breeding, especially as public plant science is increasingly abandoning this field in favour of more fundamental research. In Kock's account, the value of patents lies in their function as a backstop that prevents the well of investments and plant traits from drying up. He thinks of patents less as a fence to keep others out than as a sheet pile that channels money and incentives where they are needed. This is a position shared by his fellow patent lawyer Olaf Malek, who defended the *Tomatoes* patent at the EPO.[2] It is rare to find a patent lawyer who does not believe in the value of patents – after all, it is their job to write patent applications and get them granted.

And yet, Kock has found himself at the other side of the court room. He has opposed the *Broccoli* patent at the EPO, leading the opposition as his company's representative throughout the process of appeals. What would lead someone who advocates for patents in plant breeding, of all people, to object them in a landmark case? "Anyone, from the top of their head, would've said, this must not be granted!"[3] For Kock, the problem is that the *Broccoli* application lacks what would make it valuable after the patent's expiration: "That's why ... we said this thing is not reproducible. No deposit, no markers, no nothing. It's not inventive either, the EPO could have killed this with ordinary patent law."[4]

In other words, Plant Bioscience Ltd., the patentee, was not upholding its end of the patent bargain: for there to be positive spillovers across the legal fence, there must be something in the patent application that allows others to reproduce the invention. After all, patent protection is also an incentive to share technology in time, rather than keep it secret forever. The *Broccoli* application did not contain anything of the sorts, which to Kock meant that the applicants were not meeting their part of the bargain for the protection given by the state. Kock may be a patent advocate, but at the end of the day, he is also a patent attorney. In *Broccoli/Tomatoes II*, legal values therefore came before economic ones for him: the general benefits of native trait patents cannot be weighed against the flaws of a specific application.

All around patents: frames, hedges, thickets

Others have a less favourable view of patents on plant traits. Among them, ironically, are the ones Kock deems their first and foremost beneficiaries: commercial breeders. "Basically, this is why we always say that plant variety protection is the primary protection regime", Alexandra Bönsch, then in-house council at the Federal Association of German Plant Breeders (BDP), summarizes the majority opinion in the European seed industry.[5] "It's not immediately compensated. … You're not getting an additional incentive, but what we say is: okay, that's the haves/no-haves discussion again. Everyone has [developed something] at some point … but they can protect it as a variety." Plant breeders see Kock's point about incentives and rewards, but their calculation is a different one. Yes, everyone can use a trait in a commercialized variety to include it in their own material. PVP will prevent flat-out plagiarizing of commodified seed, but it deliberately refrains from putting fences around traits.

> 'What we say, nevertheless, is: okay, over the long run, that will even out in plant variety protection, because – yes – everyone can use your material, they can breed it into their pools; but you, as a breeder, will profit from the possibility of taking all these materials as well. And in the end we've always said that it'll even out in the end this way. And that's also the breeders' stance as far as I'm concerned.'[6]

The calculation Bönsch proposes here does not follow the logic of fixed patent terms or returns on investments, at least not when it comes to traits. For finished varieties, the BDP believes in intellectual property and exclusivity. Upstream the value chain, however, the breeders' lobby argues that such a logic would do more harm than good. Even if investments and benefits are unevenly distributed at this point in time, Bönsch points out, the differences will level over time. You might have a great trait now, but in ten years' time, you might need a resistance trait or an enzyme inhibitor that can be found in your competitor's varieties. Trait breeding, the BDP argues, is not about haves or no-haves; it's about giving and taking.

Indeed, European plant breeding has a long tradition of cooperation in competition (Brandl, 2017; Brandl and Glenna, 2017). Seed companies will behave like fierce competitors when it comes to marketing and defending their commercial seed, but they will readily join forces when it comes to fundamental research or developing new technology platforms. PVP mirrors this culture of competitive cooperation by protecting only the immediate product of variety development. In turn, it gives breeders mutual access to the fruits of their labour: plant varieties may not be copied, but they can be exploited by anyone for adding to and improving existing varieties as well as for including valuable traits in their germplasm. Some breeders, like Hubert Kempf, will even mutually share their varieties with their competitors before they officially receive market admission and variety protection.[7]

From the public point of view, the value of this arrangement lies in squaring market economics with the aim of rapid technology diffusion. Among others, this is a legacy of the post-war consensus in PVP, when politics had an interest in the overall improvement of all food crops (Bonneuil and Thomas, 2009, pp 231–267; Braun, 2021a). The faster beneficial traits travel through the commercial gene pool of a crop species, the better for society. But on a business level, the calculation is not one of private profits versus public benefits. Instead, Bönsch evokes an economics of reciprocity: the breeders' privilege to use her competitors' traits is mirrored by her obligation to share her own traits with them. It is not important when exactly she will repay them in kind: plant breeders are patient. What matters is that, over time, the difference between losing out and freeriding will disappear. "That's what I meant with time spans of 10 years or more", Carl-Stephan Schäfer, the BDP's managing director explains. "Everyone knows that there will be haves at times and no-haves at others and that in the long run they are profiting from the system. And that's why the system has been so successful."[8]

Under PVP, trait breeding shows remarkable similarities to what anthropologists call 'gift economies' (Gregory, 1982; Mauss, 2002; Bourdieu, 2008).[9] Gifts are only free on first sight: in the act of giving, the act of giving back is already implied (Bourdieu, 2008). This logic is not alien to us just because we live in a 'market society' (Miller, 2002). We feel the urge to reciprocate gifts, to not only be on the receiving side of birthday presents and buying drinks. At the same time, we hate to think about gifts in 'calculating' ways: when we turn ordinary commodities into gifts, we remove the price tag, and we make sure not to let the recipient know how much exactly they are worth to us (Braun, Brill and Dobeson, 2021; Dobeson, Brill and Braun, 2023). Likewise, the urge to repay a favour straight away or, even worse, in money, comes close to an insult (Bourdieu, 2008, pp 98–111). The removal of anything resembling a market logic, especially money as a mediator, but also the simultaneity of alienation and appropriation, is key in such a gift economy (Callon, 1998a). As soon as people are allowed to opt out of its long-term mechanisms and choose money instead of a counter-gift, this system would collapse. This is what the BDP is afraid would happen here: patents on traits would make breeders draw a 'frame' (Callon, 1998b) around their expenses and traits, seeking short-term compensation, rather than long-term reciprocation.

With drawing such a frame, the BDP's fear is not only that breeders would revisit their calculation, moving from long-term benefiters to short-term profiteers. Even before such a change of mind, the introduction of trait patents would lead to a tragedy of the anti-commons: a legal landscape fragmented by fences that protect individual interests but make the collective endeavour of breeding impossible. Just like barbed wire put an end to the free movement of cattle herds in the western US (Netz, 2009, pp 1–55) and hedges cut through the customs and practices of British commoners (Blomley, 2007), patents would interrupt the free flow of traits through the gene pool. Where the PVP fence includes cattle guards that will permit the free passage of certain forms of plants and uses, the only opening in the patent fence is a toll booth. Without permission and the payment of license fees, breeders would immediately be cut off from any trait that is not already in their nurseries. This would give patent holders an unfair advantage over their competitors: they could ask for an unreasonably high compensation or refuse to give a license altogether. In the case of a so-called 'must-have trait', without which a variety would fail on the market, such a scenario could mean the end of a seed company.

No matter how critical most European seed companies may be of plant patents, several of them have begun to build patent portfolios themselves. Some are surprisingly prolific: out of 420 patent applications for conventionally bred plants at the EPO between 2000 and 2018, a full 92 (or 21.9 per cent) were filed by specialized vegetable breeders.[10] While some of these applicants were subsidiaries of biotechnology multinationals like Monsanto, others were family-owned traditional breeding firms. Rijk Zwaan, for example, a Dutch vegetable breeder, accounts for 10 per cent or 42 of these applications. This number is only topped by Monsanto, which filed a total of 61 trait patent applications across all its subsidiaries. Other 'traditional' seed companies among the applicants include Enza Zaden (five applications) or Bejo (one application). Like Rijk Zwaan, these are Dutch vegetable seed producers. If you follow the BDP's reasoning, they are the kinds of companies that should be most opposed to seed patenting, as it undermines their business model. So why do they nevertheless engage in patenting? A head of IP working for a vegetable breeder offers yet another way for valuing patents.

'We do it. ... But the reason for that is that we saw this development by technology, more and more patents. And, also, the consolidation; bigger and bigger companies were created. Because in the 1990s we had Syngenta, there was SaatUnion still [an] independent company, and they are not completely independent anymore. We had Royal Sluis, which later, through a lot of takeovers, became part of Monsanto, Seminis-Monsanto. We thought, okay, we do not need a patent protection for the [plants], because we have plant breeders' right[s] for that. But we need the patents to protect ourselves. Because if all these big multinationals – and [our firm] is, compared to the multinationals, still small, but at that time it was really small. … For example … if a competitor would have a patent on a very important characteristic in lettuce, and we couldn't get a license, then it would mean that we miss half of our turnover. And that is the end of the company.'[11]

Like the BDP, the company disapproves of patents in conventional plant breeding. The threat of being locked out from key traits by seed multinationals has however prompted them to file for patent protection themselves. Like the economic mainstream theory of patents, they think of patents as hedge or fence that will protect them from competitors. The key difference, however, is what is inside the fence: where for Landes and Posner (2003), for example, a fence encloses and protects a valuable object of property, the fence in question here is meant to protect the company itself and ensure its survival. The patent portfolio serves as a bulwark against hostile business practices, as an arsenal that would allow the company to retaliate in the event of an attempt to lock it out.

This firm has not developed an active business model around its patents. In pure cash flow terms, it thinks of them as liability. Its patented traits are liberally licensed to competitors upon request but royalties are only meant to cover the additional expenses the company incurs by filing and renewing patents. Most of the patents fail to do so, yet the executives have decided to hold on to them for strategic reasons. In itself, this use of patents is not unusual. Next to active exploitation in the spirit of economic patent theory (Machlup, 1958), defensive patenting is a widespread practice in many industries (Lemley, 2000; Moore, 2005). Even in the plant breeding sector, it is not unusual: Rijk Zwaan, for example, openly pursues a defensive patenting strategy (Hendriks, 2017, pp 105f). Such uses of patents however complicate the narrative that patents will introduce a more economic approach to trait development in plant breeding, as argued by Michael Kock, Olaf Malek and others.[12] If the same patent can be used in two very different ways – enclosing your traits or walling your company – under two opposing patent philosophies – for or against native trait patents – what does that mean for the effect native trait patents would have on the industry 'as a whole'?

The question becomes even more difficult to answer when patents are not only used as fences to keep out others but to get a foot into the door of their operations. A patent lawyer working for another company (somewhat bitterly) remarks: "Of course we have patents. If we didn't, Monsanto wouldn't even talk to us."[13] Without a patent portfolio of their own, some breeders had to learn that Monsanto would not even invite them to discuss licensing terms for a patent. In another case, multinationals would grant a license to a firm, but without disclosing the conditions upfront. Later, it turned out, the license fee amounted to 80 per cent of the expected turnover for a plant variety.[14] There is a widespread feeling among European plant breeders – based on experiences like these – that they will only be taken seriously by biochemical companies if they are able to present an entry ticket in the form of at least one, preferably a broad portfolio of patents. This is likely owed to a corporate culture that is rooted in a 'chemical' (Brandl, 2017, p 198) or, as Carl-Stephan Schäfer puts it, 'American' tradition:[15] owing to the materialities of the agrochemical sector and the US legal context from which many of them have emerged, plant science multinationals exclusively think about innovations, products and value in terms of patents.

While some of the larger European seed producers are able to get access to this patent club, not every company can afford the costs of patent application and management. The cost of applying for patent protection in all 27 signatory countries of the European Patent Convention is estimated at €36,000 (Bently and Sherman, 2014, p 396), not including personnel costs for in-house counsel or enlisting a specialized law firm. The challenge becomes even greater for breeders when they need more than one patented trait for their varieties: obtaining a license at an exorbitant price might still be feasible as long as only one patent is involved. But paying two times 80 per cent of their sales revenues to two separate patent holders is outright impossible. And yet, the essence of plant breeding is to cross plants and remix traits. The more patented traits accumulate in one plant, the more expensive it becomes to breed. Over time, the hedges around each of these traits would grow into an impenetrable thicket of patent claims (Frison and van Zimmeren, 2021), one that only large firms with an armada of patent lawyers would be able to penetrate.

Whether they can afford entering the patent game or not, most plant breeders agree that there is only negative value in patents. Those who do not have the means to patent their traits fear a future where patent law will make it impossible for them to add value to their plants the way they are used to: by freely taking from and giving to the commercial pool of plant germplasm. Rather than protecting their labour and expenses from others, patents cut off the relations and practices that keep the seed industry alive. The few dedicated plant breeding companies that have the resources to compete in the patent race do so begrudgingly. If patents have a value for them, it is only because they feel threatened by the business practices and patent strategies of biotechnology and agrochemical multinationals whose mentality and values they do not share. In a world without patents, they would not need patents in the first place, these breeders argue. Both large and small plant breeders share the feeling that patents are forced upon them and their industry: they never asked for patents, nor has anyone ever asked them if they need them. Now, after *Broccoli/Tomatoes II*, the economic logic will squeeze many of them out of the market – or at least, that's what they fear. Perhaps, however, this is not what native trait patents are about after all.

Rent machines

It is not easy to get hold of Christoph Then. "I get so many mails", he says, "it's impossible for me to answer them all", but he eventually agrees to an interview.[16] Then is indeed a busy person. He used to be head of Greenpeace's genetic engineering and agriculture division, a function in which he opposed the Harvard University's OncoMouse patent (US 19840623774) in Europe as well as the infamous 'Edinburgh patent' (EP 0695351) which claimed a method for propagating human stem cells. In 1992, Then founded *Kein Patent auf Leben!* (No patents on life!), a network to coordinate and bundle various NGOs' and activist group's efforts against patents on life forms. Their activities do not only include legal action, public protest and lobbying against the patentability of biological materials and processes. The NGO also assists developing countries at defending themselves against biopiracy by large multinational life science companies (Parthasarathy 2017, p 137f). In response to *Broccoli/Tomatoes II*, several European NGOs and activist groups formed the coalition *No Patents on Seeds!* (NPS) headed by Then. Since 2009, he has also been head of Testbiotech, an association which seeks to provide independent technology impact assessment for biotechnology (Parthasarathy 2017, p 181f). In these capacities, he frequently appears in the media to give his opinion on the current outlook of the seed sector.

Then's explanation for the patent activities in conventional plant breeding, especially vegetables, is a shift in business models, with plant science companies increasingly taking an interest in the downstream stages of the food value chain:

> 'I guess that in vegetables the idea is to also patent the harvest. And that you don't necessarily sell seed anymore but that you get your revenues from the food sector. And that's the way that Monsanto's been leading with this broccoli and Syngenta has something like that in melons, too. And of course that is way more lucrative, to control the whole value chain and then only in the end you try to get turnover. More lucrative than, quote unquote, "just selling seed".'[17]

Then's theory of patent value follows that of Kloppenburg (2004): in plant breeding, patents are tools for extracting surplus. Their value lies in securing additional revenue across the value chain beyond the point of sale. If you do not just own the seed and the plant, but also have claims to the processed products, you can make a profit every time seed and grain change hands.

This business model is reminiscent of those for other 'novel' property objects, such as Myriad Genetic's patented breast cancer genes (Nicol et al, 2018) or Keurig's coffee machines (Perzanowski and Schultz, 2018, pp 149f). The most obvious parallel, however, is Monsanto's transgenic seed, which is sold to farmers (and only to be used) alongside its own brand of glyphosate and other chemicals. Under the agreements farmers sign, Monsanto may even retain ownership in the harvest, demoting farmers to mere service providers (Müller, 2006; Schubert, Böschen and Gill, 2011; Braun and Gill, 2018).

Where PVP carefully refrained from creating such relations of dependence and extraction between breeders and farmers (Chapter 3), patents have allowed to engineer them. The resulting object of value differs markedly from a commodity that earns its producer profits on the market. Instead, revenue is produced here by circumventing or supplanting the market with monopolies and lasting streams of income. Kean Birch and Fabian Muniesa (2020) call this process 'assetization', distinguishing it from commodification, which aims at producing market goods. Over time, commodities suffer from competition and decreasing profits, whereas assets promise lasting rents (Birch, 2020). Although patents do not automatically confer the possibility to 'ask for seconds' from consumers, they can indeed be used that way (Machlup, 1958, p 12; Landes and Posner, 2003, p 7). This happens when the legal fence used to keep out a patentee's competitors is tilted 'vertically', extending it downstream the value chain and trapping processors and consumers in a web of monetary obligations (Tellmann, 2022). Hence why some have also dubbed the relations resulting from this new entanglement of owners and owned 'neo-feudalism', as the owner's power over property extends to the people who rely on it (Morozov, 2022; see also Müller, 2006; Schubert, Böschen and Gill, 2011). The 'positive' value of the patent fence here no longer just lies in excluding others. Instead, it stems from including their valuation practices in a company's balance sheet. On the other side of the equation, this value is mirrored by the premium processors and consumers pay for a product without alternatives – downstream the value chain, one could argue, everyone else has to be worse off for this business model to work.

Indeed, many of the patent applications in the field of conventionally bred plants, among them the *Broccoli* patent, make claims to processed products, on top of traits, genes and whole plants. The so-called Carlsberg patent (EP 2384110), filed by the brewery conglomerates Carlsberg and Heineken in 2011, granted by the EPO in 2016, and opposed by Then and NPS in 2017, is another example. It claims a barley that has lost two enzymes affecting taste as well as products – especially beverages – produced from the plants. The fact that two food producers (rather than specialized barley breeders) have applied for such a patent aligns well with the scenario suggested by Then. Should barley of that type become a standard for beer, the two companies would not only hold a monopoly over the barley, but also over beer of a certain taste. Such considerations explain why Unilever felt compelled to appeal the *Tomatoes* patent – not because that patent was in itself very valuable, but because it set a precedent that might affect the whole industrial food chain from farmers to supermarkets. What is 'in' a patent, here, depends on who is in a patent: if you can turn it into an enclosure around the whole value chain, the value of the patent is equal to the extra paid by those who depend on the patented good.

As Then points out, there is considerably more money to make in the food than in the seed business: the combined sales of the top 10 companies in the seed industry totalled at US$28 billion in 2012 but were dwarfed by the US$521 billion sales volume of the top 10 food processors (Bonny, 2014, p 528). This is not surprising, given that value accumulates along value chains. Nevertheless, it shows that plant science companies' sales volumes are comparatively small and that a venture into the industries downstream could promise higher revenues and turnover. For that to happen, a patent needs to cast a web wide enough to cover all the industrial applications of a trait. The more valuable a trait after the harvest, the more likely this business model is to work – provided it can circumvent the principle of exhaustion. The key to this is replacing market transactions and sales with contracts and licenses, which prevent the property bundle from changing hands for good. Actors further down the value chain only retain a limited and temporary use right in the patented plants, forcing them to pay a patent premium for as long as the patent term lasts.

Speculative value

Then's theory is as compelling as the breeders' gloomy scenario of rapid market concentration. Still both accounts might misinterpret the value native trait patents have in the eyes of their proponents. Perhaps Michael Kock, too, is wrong about patents and their value to the seed business. Could it be that we too readily believe that they make a difference for the plant breeding industry? Sometimes a few beers help to see things with more sober eyes. After a plant breeders' meeting somewhere in a Bavarian beer garden, one of the participants lays down his alternative interpretation to his colleagues. As a seasoned plant breeder whose employer was acquired by Syngenta a few years ago, he does not buy the story of ruthless competition and profit maximizing:

'You know, we always act as if they just wanted to push us small ones out of the business, to buy us all up. You know what? They don't give a damn about us. They don't even know we exist. I'll tell you how it works. Someone in Basel has an idea for a project: "We're going to make hybrid wheat now!" So he walks up to his superior and says: "Boss, let's make hybrid wheat. With hybrid wheat, we'll … get 30 per cent of the global wheat market, which amounts to €20 billion. All I need is €20 million and 50 people." And of course his boss has no clue whatsoever of plant breeding but he really likes the idea. So he pitches it to the board of directors, who have even less a clue of wheat breeding. But all they hear is 30 per cent of the market for 20 million. So they approve it. Then the firm issues a press release: "Syngenta going into hybrid wheat, soon to hold a third of the market!" The next thing that happens is that stocks go up because of that release. So Syngenta goes to its shareholders and says "Look, you just made €25 profit on every single share you have thanks to our great business performance. Out of those €25, could you maybe give us 5 and keep the other 20 so that we can reinvest it to make you even more money?" And then the whole thing starts all over again.'[18]

Since the executives all hold company shares and stock options themselves, he goes on, this business model offers them profit along with the major shareholders. The breeding programme might be sold after five or ten years when it turns out that the whole story was unrealistic from the very beginning or Syngenta has lost interest. But by then everyone has already forgotten about it and the stock market's attention has already moved on. "And it doesn't really matter", the breeder concludes, "because everyone has profited."[19] Another breeder at the same table nods his head and agrees: "I can confirm that. I've worked for one of those companies and that's exactly how it works."[20]

This is a whole new take on the plant patent story: what if the surge of patents on pepper, lettuce, tomato or broccoli is just another business stunt pulled to collect money at the stock markets? Suddenly, all the loose ends can be tied up – the lack of market potential for most of trait patents, the rush of multinationals into vegetables and wheat, the upcoming mergers in the seed sector, even the anti-patent campaigns of breeders and activists. After all, any publicity is good publicity in such a scheme: the more the economic value of patents and their potential to radically reshape the seed business is exaggerated by their opponents, the more credible the story becomes (Vinsel, 2021). Here, credit and credibility, both rooted in the common etymology of *credo*, Latin for *belief*, rejoin (Kang, 2015).

The parallels between this breeder's account and Kean Birch's (2017) analysis of value in the bioeconomy is difficult to overlook. Observing that biotech companies' market capitalization (that is, their combined stock value) is in the billions, Birch points out that most of them never come up with a product for sale. But that does not matter as long as investors and stock markets confuse patents with products, patenting with innovating, pompous press releases with profitable projects. Suddenly, patents are counted and the quantity of applications and grants are taken as indicators for the quality of a company's R&D. Start-ups dress themselves up for acquisition (Graham et al, 2009, pp 1296–1308), listed companies for the stock market. Shareholders and investors, including the executives of plant science companies, thus take the value of patents as given, whether they have an impact on the market or not. Patents do become market objects in this over-financialized business scheme, but, as Hyo Yoon Kang (2020) remarks, they are entirely self-referential. The less people know about what is 'in' patents, the more valuable they become. Kang criticizes this reification of patents, which confuses the fence with what is enclosed by it (Kang, 2023). But because patents are highly technical, requiring legal, business and scientific expertise to properly understand them, few people can see through and properly evaluate them.

The beer garden theory of value implies that to some extent, this is a deliberate scheme. It relies on an unequal distribution of expertise and knowledge across the ranks of a firm and across the stock market. There are some stake- and shareholders who are treated preferentially – those who already have stocks and responsibility in a company, and those who are lured in to invest by hype and overblown expectations. Philipp Mirowski (2012) calls this logic a 'Ponzi scheme'. It sells promises to a gullible audience without ever delivering, instead relying on a steady influx of fresh money that will be redistributed to those who are already part. This business scheme will work as long as there is still fresh money or, in Kang's (2015) words, credit out there. Once liquidity dries up, however, the music stops and the pyramid collapses (Mirowski, 2012, p 287), leaving the last ones to invest their money in it to lose out. By that measure, the Syngenta shareholders have made the right call at the right time when they sold their stocks to ChemChina.[21] Monsanto, too, managed to woo Bayer into acquiring the company for a stellar US$63 billion in 2016 (Elmore, 2021, 267–288).

The 'emptying out' of patents is not limited to the stock markets or public imagination, however, and it is not necessarily a conscious strategy. Even within companies, patents are increasingly used as management instruments for legitimizing or 'authorizing' (Gill et al, 2012) the work of departments: an R&D department can signal its value to the company by an increasing output of patentable innovations, which the IP department will turn into patent applications, in turn justifying the expenses of both in the eyes of controlling and human resources (Long, 2002, pp 651–653; Böschen et al, 2013, p 194). Here, too, quantitative patent metrics are being mistaken for indicators of quality because the actual content of patents is not transparent to outsiders. Perhaps this is inevitable when a company exceeds a certain size, as the various departments and divisions of the organization no longer know or understand what is going on elsewhere. Patents are a means of communication to bridge these widening gaps: to an extent, they can be understood and processed by research divisions, legal departments, accounting and controlling alike.

But the more patents travel through the firm, the less defined they will become, as fewer and fewer people are able to grasp them as more-than-legal, more-than-economic or more-than-scientific objects (Star, 1989). As Christoph Then suggests, even the EPO is complicit in the inflation of patents, as it receives (and relies on) application fees but has little incentive to deny its 'customers' a grant.[22] There has also been pressure from the director's office to speed up the processing of applications to reduce the backlog (Dunlop, 2017), a policy that likely affects the scrutiny with which examiners read a patent application.

One way or another, patents show a tendency of being 'hollowed out', of becoming empty signifiers rather than working boundary objects (Star and Griesemer, 1989) that meaningfully connect and equally benefit the realms of science, law and business. Whether they are used as value signals to shareholders and markets or as proof of productivity within the company, the bottom line here is that there is nothing of value in patents after all. Like a dense hedge that closes off an alleged estate from outside view, patents hide their emptiness behind technical language and the air or innovativeness. Their real value lies in making people believe that they enclose and protect anything in the first place.

Filling the void

Is this true? Are plant patents really just a Ponzi scheme? Is the EPO complicit in the inflation of patents? Do IP and R&D departments unconsciously collaborate to pump out patents? Michael Kock's answer is straightforward: it's 'nonsense'. But do these departments not need to justify their jobs? 'Yes, but not with patents, of all things!' He goes on to explain:

> 'My KPIs [key performance indicators] were cost management. So basically, we made sure that the patents we have are actually in active use. Either in relation to products or not. My KPI was efficiency-driven. Reducing costs, keeping only what is actually relevant, and that's my KPI. So in number of patents – yes, I know, [other firms in the industry have] such a KPI, "We need to make at least 1000 patents a year," but that's complete bullshit. And no investor will buy that anymore either. So it's quality, not quantity. If I can protect my product portfolio with a smaller number of patents, then that's a considerable efficiency factor. The patent itself does not have any added value. The surplus is only what's protected by it. And if it doesn't protect, if I enclose a piece of desert, then I'm wasting money! So [that's] a very outdated vision; it's done way more pragmatically these days.'[23]

This, of course, is only half a rebuttal: Kock acknowledges that there are companies that simply count their patents, just like that there have been investors who have fallen for such a simulation of value. Yet he also insists that under his reign, and in the agri-biotech industry more general, this kind of patent valuation has become a thing of the past. Kock does not believe in patents as assets that directly translate into financial value at the stock market. Patents – and here he agrees with those who criticize their use as empty signifiers – need to be grounded in real value, that is, products. A patent which, after a given time, does not translate into a product and a market for it, must be considered worthless, Kock says:

> 'At the point of ten years after application, there mustn't be any patent that does not make any money, that's out of the question. Development takes up to ten years, but then there has to be decision: top or flop? … And then we killed them, as simple as that. We had a high degree of efficiency. It's either an active product or an active project. Or it's gone.'[24]

A patent, Kock points out, is worth its equivalent in sales. Kock's script for evaluating patents is straightforward: how much money does the company make from selling a patent-protected product, and does this sum outweigh the costs of application and renewal? If so, the patent is of value; keep it. If not, it is a liability; let go of it. To this end, the black box of the patent needs to be opened and translated into data about sales and revenues that allow a company to understand its intellectual property as first and foremost an aggregate of market prices.

For Kock, this is the only way to really grasp the value of a patent. Any evaluation from the outside, like the one offered by specialized consulting firms, "is bullshit". Patents' value cannot be estimated in isolation from the products they protect; even widely used indicators like citation numbers are useless for this reason.[25] After Kock took office as Syngenta's head of IP in late 2010, the company's patent applications and grants indeed show a marked decline until his departure in 2016.[26] This may or may not be proof of successful management – in any case, it is clearly at odds with a business strategy of printing patents.

Kock's philosophy mirrors a trend toward professionalization in IP management. Ernst (2017) evokes a contrast between old and new regimes of IP management, the former being characterized by a singular focus on patent numbers, rather technical protocols for filing and managing patents, and a weak integration in the company's overall strategy. In such a regime, IP departments are often part of a much bigger legal department and do not directly answer to the CEO. In contrast, 'modern' IP departments act independently of the legal departments, and are led by a head of IP (like

Kock) who directly answers to the CEO. In this second regime, the focus is on quality, that is, value of patents. According to Ernst, their decision to patent certain innovations largely depends on how important they are in the context of the firm strategy.

Not everything that can be patented can also be sold. "It was less than a quarter of the things that we actually managed to bring to the market", Kock says. "That's why they changed the strategy and said it's better to file late once you know it's going to be a success on the market. Doesn't make sense otherwise, it's just a waste of money." Kock's contrast foil are the plant breeders that bulk up extensive patent portfolios. Rijk Zwaan, Kock suspects, is likely the only one to have made a real profit from a native trait patent – one on lettuce resistant against *Nasinovia* lice, filed in the late 1990s and expired in 2017 (WO 1997046080A1).

Cutting back the thicket

This resistance proved indispensable for breeding healthy lettuce, an experience that likely made vegetable breeders realize how precarious their own situation was. Still, Kock stresses, it remains the exception. Together with Floris ten Have, Kock has compiled an analysis of the markets for traits and varieties in Europe (Kock and ten Have, 2016). They conclude that the *Nasinovia* patent is the only one to have had a major impact on the industry, ending up in 467 different varieties of lettuce across Europe, 255 of which belong to competitors of Rijk Zwaan (Kock and ten Have, 2016, p 503). No other native trait patent comes close to this significance: even the hybrid Ogu-Inra rapeseed developed by the French research institute INRA to revolutionize French oilseed production (Bonneuil and Thomas, 2009, pp 307–310) only accounts for 50 plant varieties affected by patent protection (Kock and ten Have, 2016, pp 503). Kock takes this as proof of the limited negative impact plant patents have had on the European plant breeding industry, although it may well prove the opposite: in someone else's hands, a blockbuster patent like Rijk Zwaan's could have obliterated the whole lettuce sector.

Kock acknowledges breeders' fears and the destructive potential of the power to exclude wielded by patent holders. Although he is convinced of the value of trait breeding and the positive contribution of patents to the industry, he believes that the right to exclude is unnecessary. He dreams of a patent law reform, which would cut back the bundle of rights that comes with plant patents. If the seed industry were to throw its combined weight behind an effort to revise the EU Biotechnology Directive for plants, he is sure that access to patented traits could be facilitated. The right to exclude could be dropped for patents related to plant breeding and replaced with a remuneration mechanism for the use of patented traits.[27] The rest of the seed sector, however, is reluctant. Any attempt to renegotiate the Biotechnology Directive, the BDP's Alexandra Bönsch fears, would be met with opposition from much bigger and more powerful industries.[28] For the time being, Kock has taken matters into his own hands. While at Syngenta, he launched Traitability, a catalogue of the company's native traits, which offers them at standard licensing conditions to third parties.

Kock's most ambitious project, however, is ILP Vegetable, a so-called clearinghouse for patents on native traits, which is meant to fix the shortcomings of plant patents with contract law (Van Overwalle, 2017b). Founded by Syngenta and ten other seed producers in 2014, the ILP members submit to self-imposed rules meant to maximize the economic benefits, which together turn patent law's right to exclude into a carefully assembled script of financial inclusion (Van Overwalle, 2017a). The rules force members to waive almost all of the rights they have been granted through their patents. Within the framework of the ILP, any member can freely breed, do research with and even commercialize other members' traits – provided it pays an appropriate license fee and makes the resulting traits available under the same conditions. The platform's rules only allow royalties to be based on net sales of the commercialized trait, thus defining patent value strictly in terms of the market product. If patentee and licensee fail to agree on a price, they trigger a so-called 'baseball arbitration' mechanism (Monhait, 2013): both parties have to submit bids, after which they get a final chance to agree on a price. If they still insist on their bids, they handed over to a committee, which has to pick the one that is deemed the fairest. The committee cannot intervene by proposing or setting an intermediary price, it has to pick one option, leaving the losing side to pay for the costs of arbitrage (Kock and ten Have, 2016, p 506; Van Overwalle, 2017b, pp 91f). This is meant as a strong incentive for members not to use price setting as a discriminatory instrument while ensuring that patent holders are adequately rewarded. In Kock's words, the ILP is about "free access, but not access for free".[29]

It is tempting to dismiss the ILP as a corporate PR bluff for selling the idea of plant patents to a sceptical public (Girard, 2015, p 14). Christoph Then, for example, is not convinced by such projects: to him, they are an attempt to calm legislators and critics into believing that regulation and legislation are not needed to solve the patent problem.[30] At the BDP, Carl-Stephan Schäfer is doubtful that the platform can bring about a change in patent culture: "It all sounds very fancy", he agrees, "I don't want to badmouth it. But someone who has patented already has a strategy in his head; he might others not to have a share of it. I know, he has made an investment, too, I'm aware of that. But by patenting a trait he can restrict the use of others. That is the main difference between PVP and patents".

With this calculation at the heart of patenting, Schäfer is doubtful the ILP will work on a long term for all species: "I don't know if such a person can be convinced to participate in such a platform." For the managing director of the German breeders' association, a private solution like the ILP is not the preferable one, because it cannot ensure that everyone will play along: "[Equal rules on the market] are, in our view, best achieved by laws, by state action. That's a bit boring, but in the end, it's like that."[31]

In a legal commentary that sums up both Then's and Schäfer's reservations, Fabien Girard remarks that 'it is still unclear why, rather than opposing the patentability of the product of an essentially biological process, the president of the EPO, along with the vegetable breeding companies involved in the ILP, keep supporting pro-patent options' (Girard, 2015, p 14). Kock's reply is that entrusting the problem to politics alone would be naïve. "I think you always need a combination of industry or stakeholder initiative and support by the legislator. I don't think we can achieve a global change in legislation", he explains. It is not an either-or but rather "working with carrot and stick, if you will".[32] In the current situation, the good-willed among the companies in the seed industry need to make the best of the legal options available to lead the way (Kock and ten Have, 2016, pp 504f).

Conclusion

Perhaps, after all, the question of value of patents is not about what subjects attach to them, nor about the objects that are enclosed in and by them. Looking for them in either place, therefore, could mislead us. It is likely that no one at the Volcani Center, the very organization that produced the shrivelling tomatoes, saw an awful lot of value in them, let alone imagined that they would end up causing all this commotion. After Arthur Schaffer, a breeder at the institute, successfully obtained them from extensive cross-breeding between wild and cultivated varieties in the 1990s, the centre filed a global patent application at the World Intellectual Property Office, which then passed it on to patent offices around the world. Nowhere did the patent end up being controversial – except for Europe (Braun, 2023).

The Volcani Center regularly files for patent protection for its plants. Not because it is out for profits, not because it needs to defend itself, not because it needs to prove anything to investors – like many similar research facilities around the world, it is simply obliged to. Established by Jewish settlers in 1920s Palestine, it was taken over by the newly founded State of Israel and named after its late director, Yitzhak Volcani, one of the pioneers of Zionist agriculture. In the first decades of its existence, the Volcani Center was there to support Israeli agriculture by developing new agricultural technology, educating farmers and introducing and adapting foreign plants to the climate of the region. Like elsewhere, however, the importance of the Israeli agrarian sector faded with its progress; food safety became less of a concern as Western countries grew more prosperous. Along with many other agricultural research centres around the world, the Volcani Center had to look for new income and purposes as the advancement of domestic agriculture alone no longer justified lavish public funding.

Since the 1980s, the centre has greatly expanded its patenting activities at home and elsewhere, licensing many of its research outputs to third parties around the world. There is no further business model behind this: the Volcani Center cannot turn its patents into commercial products itself – including the *Tomatoes* patent. The applicant of the *Broccoli* patent, PBL Technology, comes from a similar background: it is a joint venture of public and semi-public plant science facilities in the UK tasked with patenting and managing patents for them. With the reforms of British higher education and public research since the 1980s, mirrored and followed by neoliberal policy changes in other countries, plant scientists are under pressure to also legitimize their activities financially (Stengel et al, 2009; Galushko and Gray, 2013; Braun, 2023). The obligation to patent is written into funding agreements today; to make the most public value out of their research, scientists have to patent it (Slaughter and Rhoades, 2004; Mirowski, 2011). The Volcani Center and many other research institutes around the world comply with this imperative. For most public researchers today, patents are not a motivation but an obligation: they operate in an ecosystem in which patents are not up for debate anymore (Biagioli, 2019).

At the very end of his 1958 report for the judicial committee of the US Congress, the economist Fritz Machlup draws an inconclusive conclusion about the value of a patent system:

> If we did not have a patent system, it would be irresponsible, on the basis of our present knowledge of its economic consequences, to recommend instituting one. But since we have had a patent system for a long time, it would be irresponsible, on the basis of our present knowledge, to recommend abolishing it. (Machlup, 1958, p 80)

Machlup's report is a rather sobering commentary on the various attempts to justify, define or calculate the value patents have for companies and the economy overall. None of them, Machlup argues, is convincing or without contradictions – they cannot be answered independently of the patent machinery that is already in place, as Mario Biagioli (2019) remarks 60 years later. The value of patents is not simply 'inside' our 'outside' patents, in the protected traits or in the heads of company executives and economists. Where it resides and what it amounts to rather depends on how it is situated (de Laet, 2000): the same patent can be of very different value for a multinational biotechnology company and a family-owned plant breeding firm. Seed patents lend themselves to many different uses and interpretations. Depending on the context they are embedded in, they can be legal fences, enclosures, links between science and industry, deterrence, insurance, entry tickets, assets, commodities, signals, empty signifiers and many things more (Torrisi et al, 2016). The same goes for their value: it can be defined in scientific (novelty, disclosure, enabling), legal (enforceability, compatibility, validity), economic (incentive, technological progress) or business (productivity, performance, innovation) terms (Biagioli, 2019).

The problem here is that these different values and uses of patents are entangled. The defensive patenting by vegetable breeders only makes sense in a business environment in which patents can be (and have been) used aggressively. In theory, the economic value of a patent is tied to its scientific value, yet in many contexts, patents can only become valuable if they are abstracted from their scientific and technological contents. And even if everyone in the seed industry agreed to change the law and logics of patents, this cannot be done without taking into account other industries, where patents may do and mean different things altogether. The various spheres in which patents have meaning and value are not separate but interact with each other – crucially, through patents.

Over the course of their existence, the scripts bundled in patents have changed and multiplied (Machlup, 1958; Pottage and Sherman, 2010). Although from time to time the bundle is cut back and select scripts are modified, reduced or dropped (Sivaraman, 2013; Parthasarathy, 2017, pp 117–154), the short history of European trait patents suggests that the uses of patents will become more, rather than less, in the future. The trouble with being pragmatic about patents is that they are used to address so many different issues: companies, industries and economies rely on patents in various ways, hence why Machlup cautioned against their premature abolition. Even if the individual or aggregate effects of their various uses are not necessarily intended or beneficial, it is very difficult to make patents do only one thing.

Perhaps this is a lesson we can draw for property in general: if the value of property could easily be expressed in numbers or dollars, if the private profits of the owners could easily be weighed against the social benefits for the rest of society (Biagioli, 2019), the economy of property would be neither 'political' nor 'cultural'. Property would become an amorphous mass of 'wealth' that could be moved from one bank account to another; the swift calculation of need, merit or demand would save us from political debates over what and whom property is for. Alas, it is not that simple. People are attached to their property for more than financial reasons, and there are many different ways of making economic use of one's possessions.

To say that property is cultural, nevertheless, does not mean that it is inconsequential. Far from it: reining in or encouraging certain uses also has an effect on all the others. Speak about raising inheritance tax and people will hear that you are going to rob them of their grandmother's home;

suggest to sell a company to a foreign investor and a concerned public will talk of sell-out of the whole country. One reason we cannot let go of property, patents or other, is that we are attached to it in so many ways. If the multiplicity of these attachments is what makes property 'cultural', it is the lack of a shared method for weighing them against each other that makes property truly 'political'.

6

Too Much Property

It is tempting to conclude that patents are the problem. After all, this is what plant breeders have been arguing re *Broccoli/Tomatoes II*: patent philosophy runs counter to the industry's culture; patents are neither welcome nor needed in the seed sector; far from fixing the market, patents will disrupt it. Accordingly, the reaction to the EPO's decision in the case has been to try to revert it, by legal and political means.[1] It seems that all would be well if plant breeding remained removed from the reach of patent law: once native trait patents would go away, plant breeders could go back to their job of making better plants.

Not every commentator would agree. For Jack Kloppenburg (2004), for example, patents are just one part of a larger story, the privatization and commodification of plant breeding at the expense of farmers and the public. Thom van Dooren, too, argues that the category of property itself, not only patents, is at odds with plants and their way of being in the world: it runs counter to the flows of genes, practices and human–plant relations (van Dooren, 2007a, p 242). Indeed the debates about patents on plants and other life all too often frame patents as a rupture. They tacitly or explicitly assume that the philosophy of patents represents a break with the logic of other things: the uninhibited flow of goods and knowledge, the free use of our possessions, the biology of plants, the sanctity of life. The intrusion of patents into the seed sector, along with other industries, is presented as a paradigm shift. Linked to landmark cases like *Chakrabarty* and *Broccoli/Tomatoes II* or decisive legislation like the Bayh–Dole Act and the Biotechnology Directive, the discussion of seed patents is framed in terms of 'before' and 'after'.

Property issues in plant breeding are not limited to patents, however. The problem runs deeper: there are many additional forms of property that affect seed and plants at various points of their social life. While patents have received most of the critical attention recently, this does not mean in turn that other forms of property are unproblematic. Above all, this concerns PVP: all too often, it is suggested as the good 'other' of patents, as the more sensitive, custom-tailored property regime (Bently and Sherman, 2014, p 665;

Brandl, 2017, pp 208f). Where patents confuse plants with machines, inventions or a conglomerate of traits, PVP understands that they need to be appreciated as a whole while the established practices of plant breeding must not be interfered with, this argument goes (Brandl and Glenna, 2017; Braun and Gill, 2018). Neither PVP nor plant breeding, however, have remained static: they were different things half a century ago than they are today (Sanderson, 2017; Braun, 2023).

To maintain their congruence, PVP had to move on from the post-war consensus I have described in Chapter 3; a development that has not gone uncriticized (Aistara, 2012; Kloppenburg, 2014; Berry, 2023). In the 1990s, PVP underwent a revision that redefined property and ownership to keep up with the changes in agriculture and plant breeding. If the original bargain between farmers and breeders from the 1950s and 1960s could rely on widespread acceptance (at least among these two groups), the revised PVP framework was met with more resistance. In this chapter, I want to reconstruct this history as one of re-propertization, one that sought to address problems of the seed market with property. If such a market fix was successful 30 years earlier, it failed this time: rather than de-politicizing the seed sector, the extension of breeders' property in their seed led to new tensions. The attempt to push their claim to seed onto the farm was met with staunch resistance by farmers: whatever breeders were hoping to gain from the redefinition of PVP was won at the expense of their relations with farmers and the general acceptance of PVP.

Taking the farmers' point of view, Kloppenburg has rightly criticized the one-sidedness of the 1990s PVP reforms (Kloppenburg, 2014). And indeed, we could dismiss them as yet another instance of propertization as expropriation – in doing away with farmers' traditional right to reuse their seed, breeders sought to profit at farmers' expense. Framing the issue along these lines, however, is not free from its own biases. Siding with the farmers, it implicitly acknowledges their ownership of seed as legitimate, whereas the breeders' claims are rejected precisely because they violate it (Sanderson, 2017, p 231). If Chapter 3 sought to explain how PVP seed managed to become a master of two servants after World War II, this chapter asks why this seems no longer possible today. In the 1990s, property in seed started to show first cracks. The reason, as I will suggest, lies in the changing nature of farming, breeding and seed. The natural reaction of the plant breeding industry was to fill them with more property, departing from the consensus of the post-war era.

Rather than scandalizing this redefinition of PVP, I want to show how and why it has been unsuccessful. If breeders managed to expand their bundle of rights, they barely managed to add to their bundle of scripts. Conspiring with farmers, seed kept transgressing breeders' property. Realizing that the law was not enough to secure its property, the seed industry turned to biology instead. Seeking to repeat another 20th-century success story – that of hybrid maize – breeders attempted to remodel wheat, barley and other crops in ways that would align farmers' practices with the provisions of the revised PVP regime. Again, however, the property objects turned out to be noncompliant: it is possible to remake the biology of seed, but not always in a way that will satisfy breeders' expectations in their property.

While breeders are struggling to demarcate and control their property downstream, they are also finding themselves on the other end of propertization. It is not just patents that are imposed on their industry – as of lately, their use of genetic material from around the world has come under scrutiny, too. With the ratification and implementation of the so-called Nagoya Protocol, plant genetic resources are no longer 'terra nullius', a legal no man's land that can be freely appropriated. Countries of origin and local communities, long regarded as the stewards of this biodiversity, are today their owners. This is a legal headache for plant breeders in industrialized countries: if breeders have worried about their own property for the longest time, they suddenly need to think about where the traits in their nurseries come from and who they belong to.

Against the backdrop of these three cases – the extension of breeders' claims to seed, the failure to reinforce them materially with hybrid seed and the recent challenge to their sovereignty over plant genes as traits – this chapter will reflect on the 'condition' of property and its causes. Why is it that plant breeders, farmers, patent offices, developing countries and many other stakeholders run into so much trouble by applying property to seed? Although many of the ensuing controversies might resemble older political-economic conflicts about the distribution of property and wealth, I argue that the crisis of propertization is even more profound: in the face of changing materialities and shifting practices, it becomes increasingly difficult to disentangle and align the sprawling scrips of property.

Running after property: resowing fees

When peace ended is easier to answer than why it did not last. For several decades at least, both breeders and farmers were able to live with the provisions of PVP law as it was codified in the UPOV conventions since 1961 and practiced within the seed industry. Overall, the slim property solution – exclusive marketing rights but otherwise free access to seed – that PVP offered for the market problems which had emerged from the 19th to the mid-20th century worked. Towards the late 1980s, however, the first cracks began to show. Around the time of the first deliberations on the Biotechnology Directive, the European Community sought to harmonize the national frameworks for plant variety protection within its Community Plant Variety Rights (CPVR). More or less parallel to this, there was also a new amendment of the UPOV convention, passed in 1991 and entering into force in 1998. Both legal frameworks, largely developed in mutual accordance (Llewelyn and Adcock, 2006, p 494), came with considerable changes to the consensus that had so far prevailed in European plant breeding.

Among the various changes and revisions were an extended protection period, a clearer definition of the variety concept and a provisional protection during the application for PVP (Metzger and Zech, 2016, pp 14–15). Probably the most important change in the 1991 UPOV convention, however, concerned the relationship of breeders and farmers to their seed, extending the scope of breeders' property. Where the previous convention of 1978 had limited the scope of the breeder's claims to 'the production for purposes of commercial marketing, the offering for sale [and] the marketing of the reproductive or vegetative propagating material, as such, of the variety' (UPOV 1978, art 5 (1)), the corresponding provision in the 1991 version also reserved acts of 'production or reproduction (multiplication)' (UPOV 1991, art 14(1)(a)(i)) for the breeder. This meant that farmers were no longer allowed to freely save part of their harvest and resow it for next year, as growers had done for the longest time in agricultural history. Instead, they were expected to sell or process their harvest and to buy new seeds on the market.

For the longest time in the UPOV convention's history, the scope of farmers' and breeders' rights had remained unchanged: the revisions of 1972 and 1978 merely sought to extend the scope of protectable crop species and to better define the nature of the protected object, that is, the plant variety. The so-called farmers' privilege or exemption was an integral part of all previous acts up to 1991 and entitled farmers to their farm-saved seed as long as they used it for private purposes (that is, resowing) and did not sell it to others for the purpose of sowing. Although it was never explicitly mentioned in the convention's text, the definition and common interpretation of the scope of plant breeders' rights clearly left a legal space in which farmers could freely save and resow their seed on their own farms.[2] Farmers were thus free to choose if, when and how much seed they wanted to receive from the market and from their own production of seed, respectively. The UPOV convention of 1991 defined, for the first time, farmers' privilege and, at the same time, did away with it – although it offered signatories the option to reintroduce it on their own account if they deemed it necessary (Sanderson, 2017, p 234).

What had brought this change of mind about? The motives for redrawing the legal boundaries of property in seed are not entirely clear. Sanderson (2017, pp 232–238, 326) notes that there was a general tendency at the time toward strengthening breeders' rights at the expense of farmers' rights, certainly inspired by the strong protection which patents had just granted biotech firms for GM plants at the time. Nevertheless, there were also differences in the stances of the UPOV members towards the farmer's exemption. The Dutch UPOV delegation sought to retain the farmer's exemption in areas where it was already established, but wanted to prevent it from spreading into other fields where the traditional role of the farmer would not apply, for example in horticulture. In contrast, France wanted to get rid of it altogether (Sanderson, 2017, p 238). All in all, such explanations do not answer the question of what had caused or allowed for the general move towards stronger plant breeders' rights and weaker farmers' right.

An explanation for the shift within UPOV 1991 may perhaps be found in the broader development of the agricultural sector rather than in the special needs of particular markets for plants. In the years from 1961 to 1991, Western agriculture had changed considerably. Ever since the onset of industrialization, but increasingly so since the 1950s, Europe and its former colonies in North America and Australia had witnessed a strong decline of the importance of agriculture, both as an economic sector and a particular way of life. Somewhat paradoxically, however, the second half of the 20th century also saw an unparalleled professionalization and mechanization of farming, in which agricultural operations grew ever larger (Kramer, 1980; Fitzgerald, 2003). Among others, modernized farming outsourced seed processing and breeding to specialized industries, thereby abandoning the model of the farmer-breeder who produces her own seeds. In contrast, subsistence farming, access to land for cultivation and agriculture as the defining economic activity are still strong in many developing countries, which gives smallholders a different status in public opinion and political discourse (Harwood, 2012). To this day, only 53 out of 72 UPOV member countries have signed the 1991 revision whereas the rest (such as Argentina, Brazil, Kenia and Ecuador) has opted for older ones that allow farmers to resow freely. Many Latin American, African and South East Asian states, most notably India and China, never joined the UPOV convention in the first place (Metzger and Zech, 2016, p 14). This might be taken as a statement against a certain conception of agriculture, the necessity of developmental trajectories and an act of taking sides with traditional farming rather than modern breeding.

The authors of the 1991 UPOV amendment were clearly aware of this shift. In the following article of the convention, they sought to counterbalance the extension of breeders' claims to saved seed with a clear definition of exceptions to this rule – purely private purposes, research purposes and breeding purposes (UPOV 1991, art 15 (1)) – and an option for the contracting states to explicitly (re-)introduce the farmers' privilege to resow in national legislation (UPOV 1991, art 15 (2)). Despite these balancing attempts, the shifting meaning of 'private' cannot be overlooked here. A private purpose, as defined in Article 15, is any purpose not ultimately aimed at a commercial enterprise. In the terms of the 1978 convention, seed ceased to be private only if it re-entered the market as seed. In turn, anything that stopped short of marketing seed was off limits to PVP. The 1991 revision strongly restricted the meaning of 'private': under its provision, seed that hit the soil of the farm a second time (that is, seed saved and resown to produce grain) had already left the private realm.

Some contemporary observers suggested that the 1991 revision was a case of forward flight. Reporting from the deliberations, the German left-wing paper *die tageszeitung (taz)* wrote: 'What has been granted to breeders and farmers by PVP law with one hand would be taken from them with the other by patent law' (Leskien, 1991, p 9). Indeed, the few commentaries that exist from the time of the UPOV 1991 negotiations (at least in the German press) argue that breeders felt pressured by the agrochemical lobby, which sought to install patent law in the seed sector (Handelsblatt, 1987; Heinacher, 1988; Krieger, 1989; Bullard, 1990a; 1990b).[3]

In the industry itself, there exist different retrospective explanations, such as today's varieties being more thoroughly inbred and thus more long-lived; the market having become increasingly competitive; as well as a general shift in the economic landscape of breeding. Asked what prompted the change in legislation and resowing fees, however, Hubert Kempf (who started to work at the breeding station in 1991) cannot really give an explanation. "Farmers have always resowed", he points out. "I'd say it's always had the same extent. Maybe it went up a little with the [German reunification] where those giant companies resowed and there were no resowing fees."[4]

He also points to a shift in competition. Today, competition has increased five-fold compared to 20 years ago. Back then, only three to five varieties per year would be admitted to the market, and they would generally last much longer on the market:

'We ... really had the leading wheat variety in Europe in the 60s and 70s here; 'Jubilar'. And it made money for 20 years, a lot of money. Back then, you didn't have those giant companies – 'Jubilar' was simply on the market and every now and then, there were [varieties] like that. ... It was a handful of small breeders who were active and who divided the market among themselves. So if you had a big variety back in the days, it was way easier to make money over a long span of time than it is today.'[5]

Today, in comparison, economic life spans of varieties have decreased to as little as seven or eight years, Hubert says, even for excelling varieties such as 'Reform', whose outstanding performance barred any other variety of its class from being admitted in 2015.[6] Indeed, breeders' products tend to last shorter and shorter on the market before they fall out of use, as Pallauf (2018) has shown in her study of German wheat varieties. Whereas the wheat variety 'Astron', for example, admitted to the market in the late 1980s, was in cultivation until the mid-2000s and had several peaks in popularity, recently admitted varieties peak after three to five years and then quickly decline (Pallauf, 2018, p 72).

This puts pressure on breeders to recoup their expenses over a shorter span of time. Simultaneously, they have to intensify variety development to keep up with their competitors' product output. This pressure is increased by a decreasing share of wheat in overall cultivation acreage (Pallauf 2018, p 73) due to its partial replacement in the field by other economically attractive crop species. At the same time, rates of seed saving and replanting are indeed high. Although the share of saved seeds shows some elasticity, it averages around 50 per cent of all winter wheat sown in Germany, reaching 60–75 per cent in some parts of the country (Pallauf, 2018, p 76). This may seem a high number for an industrialized nation, yet it is not exceptional in international comparison. The sparse data which exist on seed saving indicate rates of 70 per cent for the US and Canada, 85 for Spain and 89 per cent for Australia in 1990. More recent estimates assume that 80 to 95 per cent of small crop seed in Australia is saved seed (Sanderson, 2017, pp 232–234, fn 8). This does not give any indication, however, that seed saving rates have been lower (or higher) in the past.

A reason not so much for the abolition of the farmers' privilege as for its introduction is provided by Carl-Stephan Schäfer at the BDP: "Back in the day", he explains, "it wasn't guaranteed that a farmer would receive a sufficient supply of certified [as in, market-bought] seed." Accordingly, the state did not want to fully entrust seed production to the breeding industry, leaving farmers with a back-up if supply of seed would fall short of demand. "This is why they said, 'Okay, I'll always have to provide the option for the farmer to resow'."[7] In pointing out the necessities of past agriculture, Schäfer also argues that they are no longer given today: supply of certified seed is anything but endangered, commercial breeders have proven over the last decades that they can reasonably provide all farmers with the seed they need, food production has not plunged into a crisis over a lack of seeds. Framing things this way, farmers' property is first and foremost a matter of societal benefits, not of private entitlement.

However, while the trends, at least in German wheat breeding, seem to be clear, they do not really explain the turn away from the farmers' privilege to resow in the 1991 UPOV convention: the 'good old times' of long-lived varieties extend well into the 1990s, as Pallauf's data show (and Hubert's timeframe of 20 years confirms). Reunification showed its effects only after the passing of the 1991 convention and even then it often took some time for them to manifest. Furthermore, all the factors described are more or less confined to Germany. In the UK, for example, the Thatcher government had privatized the largely public breeding sector at the end of the 1980s and sold many institutes and facilities to food and genetics companies (Thirtle et al, 1998; Stengel et al, 2009). Similarly, certified seed is not a universal standard in all crop species; many subsectors of the plant breeding industry (such as the tomato sector) operate differently. The UPOV revision of 1991 and its implementation in the CPVR in 1994 were international and European efforts; as such, they may only very insufficiently be described in terms of national economic developments or shifts in agriculture. Rather than providing explanations or pointing to causes, changes in the economic landscape are retrospectively referred to as a legitimization for having done away with the farmers' privilege.

While there is no traceable evidence of German farmers' resistance to the renegotiation of PVP in the press archives, this changed with the entering into force of UPOV 1991 two years later in 1993 and its planned uptake into the CPVR directive, which was passed in 1994. Perhaps more aware of the European level of policymaking than of the rather secluded UPOV meetings in Geneva, farmers' associations began to debate the implications of the end of their privilege while breeders sought to define it to their advantage.[8] The deliberations on the prospective Europe-wide protection of plant varieties however soon put PVP and the question of resowing centre stage.

The 1991 UPOV convention no longer included a general right of farmers to resow, but as already mentioned, it gave its signatories the option to reintroduce it on a national level as long as it was in accordance with and respected plant breeders' rights. The CPVR largely follows the text of the UPOV convention (often as far as down to the numbering of articles) and does not provide for a farmer's right to freely resow. One problem that emerged was that the CPVR had to account for a very heterogeneous agricultural sector. Some countries, such as the UK, had a comparatively technologized yet small agricultural sector while in others, in particular Mediterranean ones, traditional small-scale agriculture played a more important role. At the time of the passing of the CPVR, smallholders across Europe were already under increasing pressure from globalizing markets and competition by large, industrial competitors. Especially southern member states of the EC thus strongly pressed for at least a limited farmer's exemption to be included in the CPVR (Windpassinger, 1998).

The authors of the CPVR were faced with a difficult balancing act: although they sought to exempt smallholders, they also needed to prevent crafty producers from exploiting loopholes in the framework offered through an interplay between legal provisions, the different biology of crop species and the economic nature of their markets. Larger farms profit disproportionally from the right to resow, because their higher affluence and their economies of scale usually allow them to run their own seed processing facilities and to hire competent staff. In contrast, the CPVR's intent was also to protect smallholders from additional financial strain, knowing that in many cases they would resort to resowing because they lacked the necessary capital for buying new seed. The share of smallholders and their dependence on resowing, however, varied greatly depending on region and crop species. The difficulty lay in the fact that times had changed, but not everywhere, not to the same extent and not at an even pace.

The finalized version of the CPVR exempted a number of crop species from the general prohibition to resow; among them wheat, rapeseed, rye, barley, potato and rice (CPVR, art 14(2)). It stressed that the reform of the farmers' privilege was not aimed at small subsistence farmers who 'shall not be required to pay any remuneration to the holder' (CPVR, art 14(3)) and freed any area below the threshold 'which would be needed to produce 92 tonnes of cereals' (CPVR, art 14(3)) or comparable appropriate criteria from paying a remuneration. For all other farmers, financial compensation to the breeders should be 'sensibly lower than the amount charged for the licensed production' (CPVR, art 14(3)) of new, certified seed of the same variety. While remuneration was obligatory, asking for permission to resow beforehand was not; the principal freedom to resow was thereby maintained, the argument went (Würtenberger, 2014, p 107). On paper, this read like a reasonable readjustment of PVP in the European Community and a fair balancing between the interests of breeders and farmers. Farms that had left behind small-scale, subsistence agriculture should be obliged to pay for the progress achieved in plant varieties, thereby contributing to the efforts of breeders and the future of commercial breeding programmes. This only applied to seed exceeding the amount needed to plant the acreage needed for traditional family farming.

What looked fine in print nonetheless proved difficult on the ground. What, for example, did 'sensibly lower than the amount charged for the licensed production' mean? What about those farmers who had outgrown the acreage the EU considered sufficient for smallholding but still could not be considered major producers? What exactly was 'relevant information' regarding the monitoring of compliance with the CPVR? And how was the paradigm shift in property to be communicated to farmers whose degree on reliance on resown seed, regardless of business size, had reached up to 50 percent in some crop species like wheat? The CPVR left it to national laws to find answers to such questions, hoping that it would be easier on state level than across the whole EU.

Unsurprisingly, German farmers, faced with new financial burdens and no apparent corresponding benefits brought by the implementation of the CPVR, were not excited. After a four-year long controversy over the details of the new property arrangements in the seed business, farmers' and breeders' representatives settled for an average compensation of 40 per cent of the price of a variety's market price (depending on crop species). In addition, a number of rebates tied to relative seed turnover (that is, purchase of new seed) were agreed upon. Breeders had originally demanded 80 per cent, arguing that this represented the real price of developing seed, but soon found that they lacked political support for this demand (Braun, 2020). Farmers, on the other hand, had argued for a maximum of 30 per cent, but feared that remuneration would be considerably higher if they did not eventually give in to the pressure of national politics and the European Commission to settle for an agreement (Langbein, 1995; Windpassinger, 1998).

Despite this compromise, the new scheme of declaration and remuneration faced resistance and disobedience by farmers in Germany. Many of them refused to declare what and how much they had resown and a number of resowers were drawn into legal disputes. The breeders, who had tasked the Saatgut-Treuhandverwaltung (seed trust administration, STV) with overseeing and collecting remunerations, soon realized that it could not fully control their extended property in seed. While it was comparatively easy to check whether a competitor's newly admitted variety was identical to one of their own portfolio (a task that would fall within the variety offices' examination routines during the registration process) or if someone sold their seed as a generic in large quantities, farmers were dispersed all over the country. Their plots were often tiny and difficult to access, there was not even a centralized inventory of plots and seeds that breeders would have had access to (Braun, 2020).[9] So who was to go and check all the farmers' fields, bills and books to see whether they had bought or harvested, paid or pirated their seed?

Much like plant biotech firms with their patented transgenes (Charles, 2001, pp 149–191) and copyright holders in the new digital environment (Perzanowski and Schultz, 2018), German plant breeders realized that flipping intellectual property from the horizontal into the vertical dimension came with its own pitfalls. Intellectual property was originally tailored to a market that clearly distinguished between competitors and consumers, primarily seeking to fix the relationship between competitors (Braun, 2021a). This arrangement built upon implicit material conditions. Patent rights could be excessive in theory because it was incredibly difficult in practice to go from a car or a fridge to manufacturing cars and fridges (Pottage and Sherman, 2010, pp 46–64; see also Chapters 3 and 5).

The ones to feel the consequences of patent law were thus competing producers of cars and fridges, not their buyers. It is thus not surprising that patents on manufactured goods have never attracted any criticism comparable to that of patented seeds and genes. Even patents on foodstuffs were largely opposed out of fear of rising prices due to monopolies (Parthasarathy, 2017, pp 21–50), not of extended control over consumers.[10] Likewise, copyright, as an invention of the late 18th and the 19th century (Sherman and Bently, 2003), increasingly experienced crisis and public disapproval the more it was confronted with digital media whose materiality affords that users are also reproducers (Perzanowski and Schultz, 2018).

In the case of PVP, in contrast, remarkably little had changed in terms of media and objects: biologically, 1990s seed was more or less the same as 1960s seed. What had shifted, however, were economic circumstances – the modernized farmers of the 1990s were more willing to consume seed as an external input than their 1950s (or 1900s) counterparts who considered it an essential asset of the farm which had to be reproduced (Braun, 2020). In other words, farmers in Western countries could be expected to be 'ready' for consumption and willing to give up on seed reuse. Paradoxically, the authors of the 1991 UPOV convention – possibly guided by the paradigm shifts in patents and copyright – saw this as a reason to extend rather than to cut back on IP rights. Somewhat unexpected by the legislators, however, there was a considerable share of farmers who objected to this readjustment and who were all but willing to submit to a new regime of either increased consumption or increased compensation.

However, whereas Monsanto could build on an extensive network of snitch lines and contract terms (Charles, 2001, pp 149–170; Müller, 2006; Schubert, Böschen and Gill, 2011) and the digital nature of modern media allows for its copying as much as it allows for its remote control (Perzanowski and Schultz, 2018, pp 121–154) aided by respective legislation (Elkin-Koren, 2007), German breeders lacked the means to track, investigate and identify their seed. What farmers did on their private property eluded the industry. It would have taken tens if not hundreds of thousands of detectives equipped with bulky variety registers to swarm out during harvest and check the fields of farmers for varieties and their books for adequate reports to the variety holders. Notwithstanding the fact that such an enterprise would have been in crass violation of existing privacy and (land) property rights, the control apparatus would have drastically exceeded the 5,800 people employed in the German seed sector today (BDP, 2016, p 2).

The solution breeders came up with involved seed processors as checkpoints for the surveillance of seed.[11] Farmers who sought to resow their harvest needed to have it treated and prepared for storage and seeding. Compared to farms, the companies that offered such a service were few in number. Their facilities were thus geographically concentrated and an obligatory passage point (Callon, 1984) for the seed of a certain area. Seed processors could easily be addressed and, in case they did not comply but assist farmers in evading remuneration, held liable as accessories. Breeders demanded that farmers should be obliged to report variety denominations when handing over their harvest to processors and that processors in turn needed to report these back to the breeders. Furthermore, processors should report the seed they treated to the Saatgut-Treuhandverwaltung upon which the latter could take investigative action (Würtenberger, 2014), an effort for which breeders compensated them proportionally to the amount of seed processed.[12]

The CPVR had put the onus of monitoring and enforcing compliance on the breeders, while the state stayed at the sidelines. The obligation of seed processors, distributors and farmers to provide them with necessary information on request however turned out disappointing in legal practice. German courts and legislators showed limited eagerness to comply with breeders' demands by subordinating privacy law to PVP. Attempts to inquire farmers' use of seeds through nationwide questionnaires were stopped by the courts after farmers complained. Today, breeders can only ask farmers for such information if they have a probable cause to believe that the latter are dodging resowing fees (Würtenberger, 2014, p 109). Farmers, who are legally obliged to actively inform breeders of any resowing, often refuse to do so. The biggest companies, which have seed processing facilities of their own, can completely evade the surveillance system because their seed never actually leaves the farm. Professional seed processors, too, were reluctant to assist breeders with their inquiries, Hubert explains:

> 'Those big [seed processing] companies … they have mobile processing and treatment facilities which earn them part of their income. Then, of course, they're also seed sellers who are also supposed to make money from seed. So it's a balancing act for them. Because they say, "Well, if we're going to report, we're going to put off our farmers and they're going to be angry", and so on and so forth.'[13]

But if information about resowing is so scarce and insecure, how do breeders even know what they are missing out on? Can they even be sure that farmers are evading resowing fees en masse; is it perhaps not just a misguided fiction? Breeders point to fluctuations in the total amount of resowing fees in Germany. In wheat, the total volume of fees has decreased from initially €8 to 4 million before increasing again to €9 million .[14] Since the national acreage for each crop is roughly known and can be put into relation to revenues from seed sales and remunerations, the gap between what breeders expect and what they actually get can be estimated. Such fluctuations point to a shifting willingness of farmers to pay resowing fees over time. Their readiness does not just vary over time, but also from region to region:

Carl–Stephan Schäfer: It's quite interesting; in Thuringia, 97 percent of all farmers who resow actually pay up to 100 percent of the fam saved seed royalties. And then there's North Rhine-Westphalia, 23 percent … So that's the variation across Germany in 2015. These figures do not

	give any information on the actual number of famers that use farm-saved seed. It only shows the percentage of farmers that pay when using farm-saved seed.
Veit Braun:	Do you have any insights as to what kind of businesses we're talking about here? Or is that everyone from the former socialist cooperative to the family farm ...?
Carl-Stephan-Schäfer:	Well, it might reflect the size of the farms in different regions in Germany.[15]

It would be easy to dismiss farmers' evasion of resowing fees as a business strategy by large-scale cultivators, a brazen exploitation of legal loopholes and breeders' inability to control the flow of their seed. Conversely, one might defend it as a last-resort act of smallholder farmers who, hit by a bad harvest, lack the capital to buy pricey certified seed in the next year. But while part of it is certainly owed to such business strategies, Schäfer's comments also point to uncompensated resowing as a general and widespread problem. That it seemingly takes place independently of farm size and has persisted – albeit to varying extent – over the last 20 years implies that it is probably rooted not only in calculations but also convictions about what kind of property seed is and who it rightfully belongs to.

For the BDP as well as for the breeders it represents, the situation is frustrating: the state has extended their property but refuses to enforce it. Having to run the infrastructure for monitoring resowing and remunerations, they largely depend on the fleeting goodwill of seed processors and farmers. Meanwhile, they have a harder time making ends meet, with tightening competition, varieties experiencing shorter commercial lives and companies seeing less revenues from market sales. As if that was not bad enough already, the public, as far as it is aware of PVP matters in the first place, largely sides with farmers in the matter. It did not help that in 2004, right during the peak of the dispute over resowing, consumers' attention was drawn to it through an unfortunate PR stunt. Europlant, a producer of seed potatoes, pulled one of its varieties, 'Linda', from the market just as it was about to fall into public domain. The company argued that the variety was no longer suitable for commercial production due to lack of resistance and that it was trying to protect a premium segment for potato farmers from low-quality producers. Rather aggressively, Europlant sought to clean the market from 'Linda', but eventually failed because it was readmitted through the Scottish government agency SASA (Irle, 2005; brand eins, nd).

The case had little (if all) to do with resowing fees and was a matter of seed marketing, rather than PVP law. It still turned out as a disaster for the company and for public opinion of PVP. Consumers felt they were being robbed of something that was considered akin to cultural heritage at a time that cinemas were showing gloomy documentaries about Monsanto's business in developing countries. In the public imaginary, PVP, seed marketing, seed admission, GMOs and intellectual property became one big issue of farmers' and consumers' rights. In 2005, the STV was awarded a German 'Big Brother Award', explicitly mentioning 'Linda' as a prime example of its abuses of intellectual property (Tangens, 2005). If there had been any hopes for breeders to win over the public, they were certainly gone after 'Linda'.

This lack of understanding for their concerns by legislators, farmers, processors and consumers is possibly even more frustrating for plant breeders than the ill-designed remuneration scheme itself. They stress that for small breeders especially, the continued existence of their programs often depends on resowing fees: "For a big company, 9 to 10 million is nothing, for Bayer or Syngenta, that's ridiculous", Hubert says. "But if a firm like ours receives an extra 200.000 EUR in resowing fees, that's a nice thing."[16] The BDP officials in Bonn concur: "It's a lousy 25 to 30 million, altogether. Yes, it does not seem to be a lot of money, especially when you compare it to other industries". Schäfer agrees. "But for the small cereal and potato breeders, it's an existential question." "That can add up to a third of their total turnover, which comes out of licenses for certified seed, that they are missing", the BDP's press secretary Ulrike Amoruso adds.[17] It is money that cannot be invested in new breeding programmes. But as hard as such facts and numbers may be for plant breeders, they are drawn into question as soon as the perspective shifts. "Can't be that dire for the breeders", a farmer remarks during my fieldwork. "Each one of them's got a new BMW and a new Fendt tractor on their yard."[18]

"That's right", Schäfer replies. "We do have this debate in several industries: Is the work and the knowledge we have put in as well as our responsibility, that we have taken, well recognized and well paid for?" Between the lines of accounting, something else shines through here: the frustration over not being acknowledged, over one's own work not being adequately appreciated:[19] "And nowhere does it say what contribution breeders are making with new varieties, but they are taken for granted." There is a sense among farmers, Schäfer points out, that in using farm-saved seed, they are doing 100 per cent of the work themselves. But the intellectual work contributed by the breeder, he insists, is the same, regardless if seed is bought certified on the market or farm-saved.

There are alternative models for addressing the insufficiency of market revenues. The Australian state, for example, asks farmers to pay end point royalties for their harvest (Sanderson, 2017, p 242). Although there are prevailing concerns over the costs of such a system and about excessive bureaucracy among Australian breeders, this system would have the advantage of the state both granting and enforcing property rights. However, the

Australian case also shows that the state is not necessarily better at controlling and enforcing property than breeders themselves: many farmers deliberately misname their varieties as ones that have fallen out of PVP and the identity of varieties cannot easily be established upon delivery (Head, Atchison and Gates, 2012, p 59).

Both parties, farmers and breeders, claim to have property on their side. Breeders point to the law and what it entitles them to, as well as to the investments and the intellectual but also physical work they have put into seed to make it work. Without their efforts, there would not be any seed in the first place. It is the breeders' work which is mixed with the seeds and makes it grow into plants of a very specific form. From a Lockean point of view (Locke, 2003, p 288), it should therefore be their property. The breeders insist that the price for a bag of seed does not contain the resowing; if that is what farmers want, they have to make another transaction.

Farmers, in turn, are certain that the seed is theirs: have they not, after all, bought their bags of seed the same way they bought their bags of fertilizer and pesticides, often from the same supplier even? Why should anyone be able to tell them how to use what they have rightfully purchased? That the law says otherwise is a weak argument: after all, legality is not legitimacy. Farmers have inherited the right to resow and for the longest time, breeders did not seem to have a problem with that in the first place. And even so, does resowing not involve work as well? Is Schäfer's claim that the work is the same in buying as in seed saving not as ignorant of farmers' work as vice versa? From a resowing farmer's point of view (and in the eyes of a large part of the public), he and the breeder are quits (Callon and Latour, 1997; Callon, 1999) and he therefor does not owe her anything (Braun, 2020).

Legal text alone will not make property work. A solution to the resowing question that would have worked to the satisfaction of the breeders would have needed to involve far more than just the provisions of the 1991 UPOV convention, the CPVR or national PVP laws. It would have required a working apparatus for monitoring the whereabouts of seed on fields and seed bags, as well as the compliance of farmers and seed processors. Unable to provide such an extensive apparatus themselves, breeders called to the state for assistance. The state, however, considered its work done after amending the laws. Giving new property rights to the breeders was one thing; enforcing them another one, which was better left to the industry itself.

Even if the state had not been as indifferent to the breeders' concerns, it is legitimate to ask whether a working seed surveillance machine would have been proportionate to the remuneration sums in question (Gill and Brandl, 2014, pp 176–178). One way or another, the resowing controversy demonstrates how costly property can be, both in economic and in political terms. Although Schäfer emphasizes that they are not sinking money to make their point, it is also clear that resowing fees are about more than just money.[20] They are about recognition and respect, about control and entitlement. There is no legal solution to the property question here: giving the seed to the breeders is at odds with inherited rights to resow and the spirit of the market purchase (Muniesa, 2008); giving it to the farmers denies breeders' work and disrupts the market by depriving seed producers of revenues. A reduced remuneration rate of 50 per cent may look like a reasonable compromise, giving to each side what it legitimately deserves. In the end, however, it leaves both sides unsatisfied, because it may account for work but not for pride.

Going material

Frustrated by their efforts to make the resowing scheme work, breeders have turned to other solutions. While the German state has not been willing to help the breeders do their property by legal means, it is not loath to paying for private lessons in biology. And indeed, biology warrants closer engagement. Resowing is mostly an issue in cereals and potato but virtually non-existent in many other major European crop species. The reasons may partly be found in industry structures and relationships among breeders and farmers in subsectors. Breeders, however, give a different explanation: it is the seed itself which keeps farmers from resowing. Whereas wheat, potato and barley readily lend themselves to being resown, maize, rapeseed and rye refuse to collaborate with farmers who want to reuse their seed. The reason lies in different genetics and breeding systems. While the inbreeding method in wheat described in Chapter 3 used to be the standard one for most crop species in the early 20th century, many have switched to a different system today: hybrid breeding.

Unlike so-called inbred or line varieties, hybrid varieties achieve genetic and phenotypical uniformity not directly by inbreeding, but by a controlled cross-fertilization of two inbred lines. These two lines, one acting as a 'mother' (pollen recipient), the other as a 'father' (pollen donor), have been extensively tested for compatibility and inheritance pattern. Their offspring will exhibit predictable, stable characteristics over a series of cross-fertilizations. What is sold and used as seed for agricultural production is this first filial (or F_1) generation. Whereas inbred line varieties are phenotypically stable because they are genetically homogenous – meaning they possess two copies of each allele – F_1 seed has two different alleles per gene. Its uniformity relies on the balance between dominant and recessive traits: if the offspring inherits a dominant allele from its mother and a recessive one from its father, it will display the dominant one. If such F_1 seed is selfed, however, alleles will be reshuffled and two recessive copies will end up together in many cases, thereby producing new phenotypes (Acquaah, 2012, pp 131–145).

Since these phenotypes are not only unpredictable but also quite diverse across a population, this effect is highly undesirable for farmers. Plants might differ in such crucial characteristics as height, pest resistance, germination or ripening, making it difficult to cultivate them in modern industrial agriculture. Every modern agricultural process from sowing to harvesting to processing ultimately depends on the uniformity of seed and the plants it produces (Kloppenburg, 2004, p 117). This uniformity, however, disappears when a F_1 plant is selfed and resown. Farmers thus have a strong incentive not to resow but to buy new F_1 seed each year. As Berlan and Lewontin (1983, cited in Kloppenburg, 2004, p 97) have put it, hybrid seed is rendered not biologically, but economically sterile. It can only be fully appropriated once, after which the perfect fit between soil, seed and machinery dissolves.

In many crop species, hybrid breeding has given a biological answer to the resowing question. Where the limits of legal texts and of states' ability to enforce property right have become apparent, going back to the hard materiality of property objects promises a solution. Anthropology is thus not mistaken when it argues that property is about relations between people; it is just that these relations are not necessarily entrusted to social groups. They can just as well be folded into and articulated through property objects by reframing them materially. Hybrid seed becomes a vector and arbiter of property, reshaping the trajectory of seed even after sale. Technology, one cannot tire to quote Latour (1990), is society made durable. After the sociology of plant variety protection has proven so complicated and impracticable, technologies for protecting plant varieties are on the rise. Tired of breeders' complaints and apparently convinced by the effectiveness and simplicity of hybrid varieties, public and private researchers have boosted their effort to make hybrid breeding work in an increasing number of crop species (Bonneuil and Thomas, 2009, pp 310–312).[21]

Along with other German breeders, Hubert and his company have joined a research project directed by the Landesanstalt für Landwirtschaft (State Institute for Agriculture; LfL), a public research facility, aiming to bring hybrid breeding to wheat. The aim is to transform wheat from an 'inbred' into a 'hybrid' crop, with the LfL in charge of the science and seed companies providing experimental plots and elite wheat lines. The aim of the project is to develop a biological system for producing hybrid wheat seed. There are three different types of hybrid systems: mechanical systems, which work with manual or machine-assisted sterilization of parent plants; chemical systems employing sterilization agents; and biological systems which mostly rely on cytoplasmic male sterility (Becker, 2019, pp 281–285). Regardless of the mechanism used, they all seek to render the mother plant's male parts sterile so that only the father's pollen will be able to produce offspring. In absence of sterility, the mother plant would also self-fertilize, leading to a high proportion of inbred offspring, which in this case is not desired.

In maize, where hybrid breeding was originally developed, the plant's anatomy facilitates mechanical sterilization: male flowers, the so-called tassels, grow on top of the plant and are large enough to simply be cut off even by unskilled workers. In many other plants, however, flowers are much more delicate and house both female and male parts. Manual sterilization of wheat is an incredibly difficult and demanding task, which makes this species unsuitable for mechanical systems. Since the 1970s, especially US companies have therefore put their hopes in chemical agents that will sterilize mothers (Kloppenburg, 2004, pp 123–126; 242). These so-called gametocides can be sprayed on plants to inhibit the development of anthers and to produce purely female flowers (Acquaah, 2012, p 135). Many of these compounds do not work as selectively as desired, however: they either do not kill enough flowers (leaving maternal anthers intact) or too many of them (also destroying female parts). The few agents that produce satisfactory results come with serious side-effects for human health, which has led to their ban in Germany and forces hybrid wheat producers to bulk up their seed in neighbouring France, where regulations are less strict (Becker, 2019, pp 281f). Somewhat more important from a proprietary perspective is that the only currently available gametocide is subject to a patent held by the company SaatenUnion, giving it a quasi-monopoly on chemically produced hybrid wheat.

Current efforts thus focus on so-called cytoplasmic male sterility (CMS) systems. These systems build on a genetic incompatibility between the cell nucleus and the cytoplasm, which prevents plants from developing male flowers, rendering them male sterile from the seed stage on. This is usually achieved by crossing two different but closely related species: one of them will pass on its cytoplasm, which will be inherited by the offspring and prevent them from producing pollen. Such systems were first developed in the late 1930s in the US and promised a wide application across all crop species (Acquaah, 2012, p 356). While there have indeed been successful implementations (most notably in rapeseed), many major crops, among them soya bean, cotton and wheat, have proved challenging for this approach (Bonneuil and Thomas, 2009, pp 310–312). To date, one of the key difficulties in wheat is the reversal of the mothers' infertility. While their male sterility is desired for F_1 seed production, it represents an obstacle for the reproduction of the female parental line: fathers and mothers, too, have to be reproduced for a breeder to be able to produce the same F_1 hybrid seed in the next year. Whereas in inbreed varieties, variety development, production and reproduction are one and the same thing, hybrid systems pose the challenge of managing all three tasks independently (Becker, 2019, p 291).

For the LfL, the task was to find genes in spelt (*Triticum spelta*) and other relatives of wheat that are able to overcome the cytoplasmic barrier and restore fertility in the F_1 generation, allowing it to produce grain. If fertility is not restored, plants will not produce grain on farmers' fields, as they will simply inherit male sterility from their mother, resulting in empty ears.[22] This in turn leads to low yields and infection with ergot, both of which are feared by farmers. CMS systems thus depend on potent restorer genes able to at least partially reverse the effect of the incompatible cytoplasm. The LfL researchers were hoping to find at least two (ideally more) spelt genes that boost male fertility. Although results in spelt looked promising at first, the effect could not be reproduced in wheat despite all efforts, leading the researchers to conclude that there was no unknown second restorer gene. Instead, they speculated, the effect was likely due to variations across crop species.

The story highlights the fragility of hybrid seed as a technical solution to the property problem in wheat. Hybrid systems look neat when they are presented as schemes in plant breeding textbooks like Becker (2019) or Acquaah (2012): there are maintainers, restorers, male sterile lines and in the end there is hybrid seed. The further away from the textbook and the closer to the field, however, the more complicated they become. There are, for example, not only restorer but also modifier genes, which will not restore fertility by themselves but are able to boost true restorer genes. The difference between the two can only be established in large-scale field experiments. Restoration may work better in some crop species than in others; it is not a binary effect but can vary in results from one grain to a full ear. The identity and total number of restorer genes is unknown, and so is the answer to the question of how far away researchers are from a commercially viable hybrid system.

Even if breeders and scientists manage to find enough restorer genes to fully restore fertility in F_1 seed, it is doubtful whether a switch to hybrid wheat is going to happen. When hybrid breeding was introduced in maize, the decisive economic argument was that F_1 seed was more viable, healthier and above all more high-yielding than its parents. Maize, as an open-pollinated crop, had never taken well to inbreeding. With hybrid breeding, however, it could realize its full potential: the greater the genetic distance between the parents, the greater the increase in yield of the F_1 generation over the average of the parental lines. The exact extent of this so-called heterosis effect or hybrid vigour in maize is difficult to establish from the literature. Acquaah (2012; without providing a methodology) accredits 60 percent or 2400 kg/ha of yield increase in US maize between 1930 and 1970 to it. However, although other crops have successfully been hybridized and show hybrid vigour, none of them exhibits a heterosis effect remotely close to maize. In hybrid wheat, heterosis is currently estimated by breeders at around 10 per cent, but even this may be an overestimate: in rye, a genetically more suitable, obligatory open-pollinated species, it is a mere 18 per cent (Nickl, Hartl and Herz, 2014, p 35f).

The magnitude of the heterosis effect matters deeply for hybrid breeding. Hybrid seed is expensive in development and production compared to inbred seed: breeders need to develop mother and father lines independently, maintain two separate and genetically distant breeding pools, test their hereditary characteristics against each other and develop maintainer and restorer lines for each new hybrid variety. Only one of three lines – the mother line – will produce commercial seed while the father line's yield is worthless and those mother lines which have been pollinated by a maintainer only produce next year's mothers. Producers have to plant mothers and fathers in separate rows so that only the mothers' yield will be harvested. All of this adds to production costs, which the researchers at the LfL estimate at four times those of producing inbred varieties.

Accordingly, hybrid seed is more expensive, a disadvantage it has to compensate for with higher yields. If wheat hybrids do not yield about 6 per cent more on the field than their inbred competitions, farmers will not buy them.[23] Carl-Stephan Schäfer emphasizes that there will be no working hybrid system if there are no economic advantages for both breeders and farmers and that such a system cannot be forced upon farmers. "The system does not work simply because the breeder wishes so. ... The key point is that the system – yield, resistances, quality – is better for the farmer." So far, he points out, hybrid systems have been successfully established, "not because the breeder wanted it, but because it had a greater value for the user, the farmer".[24]

There is a further aspect which might be more worrying for breeders: hybrid breeding, by privatizing varieties down the value chain at the seed market, also privatizes them further upstream in variety development. Hybrid varieties are subject to the breeder's exemption and may be freely used for crossing. But since they do not genetically fall into the widely used parent gene pools but in between them, breeding them with his own material would make little sense for a breeder. Parental lines, in turn, are not usually admitted to the market. Seed companies treat them as trade secrets, which means they are unavailable for varietal development outside the firm. One result of hybrid breeding systems on the breeders' side is increased secrecy and less collaboration, favouring larger companies with bigger private gene pools.[25]

For the time being, the future of hybrid wheat is, in any case, uncertain. Many wheat breeders are not convinced that hybrid wheat is better than their current breeding system, admitting that they are 'in it because we're more or less forced to'.[26] When I come back to the LfL roughly a year later, I learn that the seed companies participating in the project are reluctant to go ahead. Syngenta has just experienced a disaster with one of their hybrid barley varieties, forcing it to recall all the seed and compensate its customers. "If I look at the sum they paid", one of the breeders is quoted, "that's ten times my turnover. I can't take that risk."[27] With the small breeders pulling out and with hybrid wheat's lack of economic advantage, an industry-wide transition to hybrid wheat seems unlikely.

"I don't think it's going happen anytime soon", Michael Kock, the ex-Syngenta manager, says. Performance is simply too weak, gene pools are not wide enough, farmers and the public will not be convinced: "There's so far simply no good solution for building sustainable systems for value creation."[28] After an initial hype for hybrids, barley breeders are slowly returning to inbred varieties. In wheat, hybrid breeding may be dead before it even started. The meagre yield surplus does not justify their continuation, especially as it is not reliable and predictable across years. Despite its drawbacks, inbreeding is still cheaper and less risky for breeders, and without a competitive advantage of hybrid wheat, inbred wheat is unlikely to disappear from the market (Braun, 2020). With regard to previous French attempts to transition to hybrid wheat, Bonneuil and Thomas (2009, p 310) speak of a 'technology bubble': time and again, the industry has made an effort to push hybrid systems for wheat, yet so far, they have failed to materialize.

Property's technology is not any more straightforward than its sociology. Reframing seed, while offering a solution to the issue of uncompensated resowing, is far from a shortcut. First of all, it is a matter of material. Is there a working mechanism for introducing and reversing sterility? Can researchers find enough restorer genes? Will basic mechanisms be translatable across species boundaries? Are restoration and heterosis strong enough to produce sufficient yields? And will the system work reliably enough for commercial purposes? Tied to such material issues, there are economic aspects: can heterosis not just compensate for costs, but also outcompete inbred varieties? Will small breeders be able to afford it or simply stick to inbred lines? Can they cope with the economic risks of failure of hybrid seed? Then of course, there are wider social implications that link back to economic and material ones, such as the privatization of R&D that seems to go along with the privatization of varieties or the prospect of fragmentation of gene pools. Material technologies may appear more durable, more powerful than negotiations and lawsuits (Latour, 1990). But they come loaded with issues of their own, which are riddled with economic, legal and social questions again. If materiality is ever a way out of the shortcomings of property in seed, it is also a way into new problems of property.

The property of others

At first glance, the Convention on Biological Diversity (CBD) is a rather dull piece of legislation. It was signed during the 1992 UN summit in Rio de Janeiro. Aimed at responsibly governing and protecting the world's biodiversity, it contains two follow-up protocols, named after the cities they were signed in: the Cartagena Protocol on import, export and trade of GMOs, signed in 2000 and entered into force in 2003, and the Nagoya Protocol on natural genetic resources, which was signed in 2010 and became effective in 2015 (Parance, 2021). As unimpressive as it may appear to outsiders, the CBD represents a major paradigm shift and a landmark in the political thought of genetic resources. Whereas before 1992, biodiversity was considered 'natural', that is, outside the boundaries of politics, the Rio summit addressed it as a political matter. Unregulated human activity had brought about an unprecedented loss of biodiversity that had to be halted by political action. But how to transform biological resources from something that was 'simply there' (but under threat of disappearance) into something that could be effectively addressed and handled by politics? The negotiating parties in Rio decided to turn biodiversity from what would have previously been considered common heritage of humankind (Hayden, 2003, p 25) into national property (van Dooren, 2007a, pp 101f).

The significance of this act cannot be underestimated. The CBD was not just intended to put a stop to uncontrolled exploitation of biological resources and their rapid demise. In the same breath, it sought to lay the foundation for reversing such activities and turning them into something productive and beneficial for the environment. In its preamble, the CBD, '[a]ffirming that the conservation of biological diversity is a common concern of humankind', also stated 'that States have sovereign rights over their own biological resources' and added that they 'are responsible for conserving their biological diversity and for using their biological resources in a sustainable manner' (CBD, 1992, p 1).

In framing biodiversity in terms of biological resources and by assigning responsibility and sovereignty to the state, the CDB established a property relationship between nation states and so-called indigenous and traditional communities on one side and biological resources on the other. The active part of this pair – the nation state and the communities – should actively exert its property rights through and ultimately for the benefit of a passive object – their biological resources. Sovereignty here did not just mean 'legitimate rule' over subjects or delegation of political agency in a Hobbesian sense. The CDB also made it clear that there was a link to establish between political authority over and commercial use of biodiversity (Heeren, 2016) by '[a]cknowledging that the provision of new and additional financial resources … can be expected to make a substantial difference in the world's ability to address the loss of biological diversity' (CBD, 1992, p 2).

This last conviction of the authors and signatories of the convention was later taken up by the Nagoya Protocol, which sought to provide clear rules for exchanging biological material, be it living organisms, tissue or DNA. The background for both the Nagoya Protocol and the underlying provisions in the CBD were increasing dissatisfaction and protest within and among African, Asian and Latin American states over the common practice of foreign biotech and medical companies venturing into biodiversity hotspots to find hitherto commercially unused substances, genes and materials which they could turn into (often patented and trademarked) products (Hayden, 2003, pp 25–47; Dreier, 2023, p 791).[29]

Activists accused these companies of biopiracy, that is, taking biodiversity, which was often embedded into traditional practices of hunting, foraging or agriculture, and turning it into commercial products and profits without making any contribution to their development and sustenance over countless generations (Dutfield, 2004, pp 1–13). Nowhere was this argument as evident as in agriculture, where ancient landraces and cultivars depended on the continuous cultivation by smallholders, many of them indigenous and often too poor to afford modern high-yielding seed (Bertacchini, 2008, pp 3–4).[30] Indeed both the *Tomatoes* and the *Broccoli* patent rely on genetic material from wild crop relatives of the two plants, collected during expeditions (Braun, 2021b; Braun, 2023). The accusation was all the more convincing as right at the time of the Rio summit, breeders' associations and UPOV were busy expanding the realm of (strengthened) PVP to exactly these countries (van Dooren, 2007a, p 31; Sanderson, 2017, pp 81–112).

The Nagoya Protocol promised to solve the problem of biopiracy by making the contribution of indigenous people and developing countries to biodiversity visible (Hayden, 2011) as well as by giving them the right to decide whether it could be used by others and entitling them to a share in the profits made with those biological resources (Heeren, 2016; 2017; Relly, 2023). In its approach, it follows a classical solution proposed by Ronald Coase (1974), who argues that the adequate reaction to uncompensated contributions to the economy is to turn these contributions into property (Bertacchini, 2008). By handing control over to the authors of these contribution, Coase explains, they are given the ability to negotiate an adequate compensation and find appropriate solutions. This allows for flexibility but also for autonomy: actors are not dependent on public subsidies or supplementary payments set by someone else but can actively make use of markets and enjoy economic freedom.[31]

The bottom line of Coase's property theory is that the only law economic actors need from the state is property law – the rest they will be able to sort out themselves. What for Coase is a case against the state as a regulatory instance makes sense in the context of international law where there is no sovereign who could dictate just prices or solutions. For many emerging and developing countries, the Nagoya scheme promised to put a stop to biopiracy and to open new sources of income. Consequently, states like Brazil hailed the adoption of the protocol by 107 parties, including the EU, in 2010 (Aubertin and Filoche, 2011, p 61).

When the Nagoya Protocol entered into force in Germany in 2014, however, many actors who had never set foot into the rainforests of Brazil suddenly realized that they were threatened by the passing of Nagoya into national and European legislation. Ecologists, natural history museums, zoos and many other non-commercial professionals and institutions realized the extent to which they are reliant on foreign genetic resources (Kupferschmidt, 2018). Even if they try to seek approval with the authorities beforehand, Nagoya is often badly institutionalized on a national level. This leads to confusion over which agency or bureau is in charge of overseeing applications, permissions and compliance. Breeders, the BDP's Alexandra Bönsch explains, are faced with similar challenges:

'There's [the countries] that respond quickly. Those are usually the ones without any [access and benefits] legislation, so for example England or the USA. ... And then there's those which respond, for example Spain: "Well ... we haven't quite implemented all of this yet, but you still have to fill out this form and send it to us. But you also have to ask the local authorities." And then you learn that, first, this form you're supposed to fill out is in Spanish and that there's no English translation. And, second, that no one really knows which regional authority is in charge and of what exactly. [So we said to our local project partner,] "Please go and see if you, with your command of Spanish, can solve this issue locally." And if we can't sort it out, then we're not going to survey that region. Simply because we can't risk that at some point ... something's going to pop up; "well no, you were never allowed to use that", and then we have to sort out that mess. Research is going to ask, "What about the findings we have generated with that?" Our breeders, too, who have maybe developed that material over several generations or even commercialized a product, are then confronted with the question of what to with it. That's impossible. We simply cannot risk that.'[32]

Insecurities over access to material and its use are one concern. Another one is the paperwork that Nagoya brings about for breeders: for everyone dealing with genetic resources, it creates an enormous bureaucratic burden. It is not just permission forms that have to be filled out; material also has to be documented so that its circulation can be monitored. If authorities announce an inspection, a company's books need to be able to account for all foreign material and its whereabouts in their varieties and breeding lines. If a breeder orders a sample from a seed bank and breeds it into her material, the resulting offspring, too, will fall under Nagoya legislation. While it is only Nagoya material that has to be documented, unclarity remains with respect to non-Nagoya material. Will inspectors accept if you document the foreign 20 per cent of your germplasm but not the other 80 per cent, explaining that the latter does not contain any relevant genetic material? To avoid any risks, breeders would have to document not only their Nagoya-relevant material, but effectively all of their other germplasm and gene bank accessions, too, to prove that these are free from foreign material.[33] Like patents, one major problem with Nagoya is the transaction costs and additional manpower it demands of companies.

The other problem with Nagoya material, as an embodiment of national property, is that it lives on and proliferates. There is no cutoff point for property claims, like the creation of a new variety in PVP or exhaustion through commercialization in patent law. There is not even a point in time after which property will cease – while PVP lasts 25 to 30 and a patent 20 years, Nagoya is, by design, eternal. Worse even, property is not restricted to the initial sample or specimen taken from a country but passed on to all future generations of plants that possess even the slightest genetic trace of Nagoya material. One way of grasping the nature of property in the Nagoya framework is to understand it as infectious property: the more Nagoya material is used, the more material will fall under its scope, despite the initial contribution of the countries of origin or their indigenous people becoming smaller and smaller in proportion.[34] The result is a dizzying sprawl of claims, especially if more and more Nagoya genes end up in a shared gene pool. Nagoya even infects, at least when it comes to practice, material that has not been bred with: only by subjecting their own material to the same bureaucracy as foreign material, breeders can be sure to upkeep the division between both.

As a reaction, breeders resort to radical quarantine. Hubert Kempf tells me that he has stopped using any material collected before 2015, the cutoff date for the Nagoya Protocol in Germany, to avoid any Nagoya germplasm entering his breeding program. Many companies have done the same. Bigger breeders, while able to afford the Nagoya bureaucracy to some extent, are none the happier about it. One company has two lawyers exclusively dealing with Nagoya issues but is nevertheless discontent with the situation: the manpower, it complains, would be better invested in breeding itself.[35] Few things unify the seed sector across all lines of division like Nagoya does. "The patent system has, in practice, never led to a blockade", Michael Kock says. In contrast, "already now, Nagoya has produced a virtual blockade. A lot of people only take the genetics they already have in their drawers and don't access anything new". This is the result of the quarantine that breeding companies and research institutes felt forced to put themselves in. If no exotic material from biodiversity hotspots is bred into commercial germplasm anymore, pest resistance and other traits are going to erode sooner or later. What is worse, Syngenta has invested several million Euros into its Nagoya compliance system, a sum smaller companies could never afford. "This system, if it practiced the way it is intended, is going to drive consolidation of the industry", Kock soberly concludes. "And this time, driven by NGOs."[36]

Kock blames activist groups, such as the Berne Declaration and the European Green Party, for having actively worked toward an implementation of Nagoya that is as strict as possible and that now fails to meet with the demands and needs of the breeding industry. Out of the four committee representatives tasked with drafting the EU directive for implementing Nagoya, all four were members of the Green Party. The other parties in the European Parliament left it to them to work out the legislation, Kock criticizes, probably assuming it was some kind of "tree-hugger law" and that biodiversity was a matter of herbs in some backyard. "The result", he concludes, "is a law with such a massively negative impact that it will be difficult to compensate over the medium term."[37]

There is a general sense throughout the seed industry that the way the EU and its member states implemented Nagoya is deeply flawed; efforts to fix the problem and find a way out of the situation should thus focus on national governments and the decision-makers in Brussels and Strasbourg. Beyond implementation on EU and national level, however, there is also a deeper dismissal of the Nagoya framework. Some suspect that the money will never even reach indigenous communities, instead ending up in the pockets of politicians and officials as yet another failed development scheme.[38] Hubert explains that he would rather give his varieties to the countries of origin for free or to only pay once a variety is commercialized, but not when he is still uncertain about the usefulness of a genetic resource. Others, like Kock, criticize Nagoya's property architecture as completely misguided. Right at a time when the industry is on a good way to fix the shortcomings of the patent system, he complains, they reappear, only this time from the side of Nagoya – not driven by the industry, but led by NGOs and environmentalists.

In principle, everyone agrees that there should be a system for access and benefit sharing. It is just that plant breeders in Europe would like to have it on their own terms: with minimal bureaucracy, clear definitions, rather a flat rate than payments tied to individual resources and sequences, with a research exemption as broad as possible. While some countries, like Brazil, have already incorporated some of these elements in their legislation (Aubertin and Filoche, 2011), the general trend is going into a different direction: Nagoya contains several loopholes, among which the blurry line between the material and the immaterial is a major one.

Collectors need to seek permission before exporting biological and genetic material, but under the current terms of the protocol, they could simply screen DNA samples in situ and just take the information. Back home, they can then synthesize the sequences from scratch without any material ever leaving the country of origin. Concerned about this new form of immaterial biopiracy, a number of countries are currently trying to close this gap by claiming ownership not just of material but also of the information contained in it. "On the one hand, that's understandable", says Bönsch, "because the idea is that at some point all you need is gene sequencing and then you can – with new plant breeding techniques or whatever – reconstruct it and won't even need access anymore."[39] But the administrative burden would become even more crushing, up to a point where the whole system would collapse under its weight.

Ideally, Nagoya should create more legal security by defining the terms for access and benefit sharing. By applying these terms, one side – the industries of the Global North – has learned that Nagoya has the opposite effect, creating more insecurity for them. Meanwhile, the developing countries have become aware that their property is imperfect and seek to fortify it even further, which will only worsen the mutual discord. The problem is not simply that property under Nagoya is ill-defined, but that it has been over-defined from the very beginning (Pottage, 2006; Dutfield, 2017). The CBD and Nagoya are more than the extension of Western ideas of property to other places in the world (Strathern, 2022, pp 188–190): as if the wave of propertization had rolled back from foreign shores, they rather represent a return of these ideas to Europe in a different guise.[40]

It comes as no surprise then that the Europeans want to move away from property in genetic resources, favouring very simple end point royalties or flat rate systems.[41] Michael Kock dreams of an even simpler solution: a turnover tax for breeding companies of 0.1 per cent, the revenues from which are then used to fund development and conservation. "Or, better even: capture the whole value chain and increase VAT for products based on biological resources by 0.1 percent. Because", as Kock explains, "only 5 percent of the value is added on the level of the breeder and the remaining 95 percent are generated down the value chain."[42] These are some wide-reaching proposals, which Kock admits are probably not going to become reality any time soon in the current political climate. They would represent a radical turn away from the self-reinforcing introduction of ever-new property on material, genes, sequences and information. Instead, they would un-define property, making it easier for everyone involved to share material and receive compensation.

Visionary as Kock's proposals may sound, it is indeed doubtful whether they would be able to address the issue more adequately. Enrico Bertacchini (2008) argues that such a 'Pigouvian' (after Arthur Cecil Pigou (1932)) approach of covering externalities by taxing and the Coasean approach embodied by Nagoya are two sides of the same coin. They both imply that at some point, there will be private profits: in Pigou's framework, they occur on one side, at the expense of social welfare, but can be taxed to remedy the social deficit they produce. Under Coase, both sides agree on a market price for social costs, allowing for profits on one side and at least a break even on the other. In the case of seeds, this entails that at some point of the value chain, seed is turned into some form of property. Bertacchini points out that the ideas behind the CDB and Nagoya did not come out of nowhere: they were a reaction by developing countries to attempts by industrialized countries to extend plant breeders' and PVP rights globally (Sanderson, 2017, pp 53–58). Establishing rights in domestic biodiversity were thus more than just an attempt to get a monetary piece of the cake. It was also a way of underscoring that the modern, high-yielding varieties of industrial agriculture depended on the efforts and co-creations of those who were yet to be modernized and often belittled for their backwardness (Dreier, 2023, pp 793–798).

Claiming property in landraces and ancient cultivars was an act of claiming authorship (Gill et al, 2012) for this contribution, of making the 'primitive' plants of the South visible alongside their people. This is an observation Alexandra Bönsch has made herself:

Alexandra Bönsch: One of the key issues, which I didn't realize in the beginning, is: at some point, the Ministry of the Environment told us, 'This is, first and foremost, a matter of self-assertion. Self-assertion of emerging and developing countries against the industrialized world; to be attributed a position and that they receive a fair compensation for it'. That's why they have difficulties accepting a multilateral system, because these property contributions are not individually accounted for. And I guess it was more about really staking a claim here.

Veit Braun: About recognition.

Alexandra Bönsch: About recognition. ... And this is the point, actually. It's a deeper level ... And we have to ask ourselves whether there's other ways of acknowledging that without pursuing this particular way, from which I am sure no one is going to profit at the end of the day.

Property may be about money here, but that money itself a token of respect and recognition rather than a source of profit: to have something is to be someone (Fromm, 2005). Kock's vision of a flat tax on valuation, simple and elegant as it may appear, will not be able to satisfy the legitimate demand of being recognized and treated like a respectable business partner, rather than a recipient of yet another form of developmental aid. In a flat-tax system, Bönsch realizes, individual contributions and authorship are swept up and erased by the flow of anonymous money.

This holds true for both sides. Breeders might complain about being left in the cold by the state when it comes to remuneration, but the prospect of a tax-based solution to their problem is not viewed as a desirable solution to their problem. When I mention the option of tax-funded, competitive bidding for trait and variety development, Carl-Stephan Schäfer stresses that this is not what breeders want. With public money, breeders fear, there is also the danger of the state setting the agenda in trait and variety breeding. Their objection is not just of an economic nature: it demonstrates a sense of pride and love of autonomy that is rooted in private property. Public funding would solve the monetary problem of resowing and seed saving, but not the property problem.

Conclusion

To ask who should rightfully own seed – farmers, breeders, traditional communities, countries of origin – implies that the problem with property lies in its distribution. Much of the discussion about farmers' rights and genetic resources indeed revolves around these important questions. The remedy, it is implied, is to redistribute property: where seed, genes, varieties and traits are claimed and taken from others at their expense, damages need to be paid or property needs to be restituted. In doing so, however, we also reify property (Nichols, 2020): it is because farmers' ownership of seed is overridden by breeders' rights the former need to be given back what is rightfully 'theirs'. Like the critique of patents, this political economy of property relies on implicit hierarchies – between those who claim property but also between different forms of property. We assume that with propertization, property will end up in the hands of the powerful. Similarly, we expect new property rights to override existing ones; that patents will trump PVP, while PVP will override farmers' ownership and traditional rights (Kloppenburg, 2004; 2014).

The cases of resowing fees, hybrid seed and Nagoya rights complicate this idea. Plant breeders may have expected their claims to outweigh those of farmers, yet it is not clear what hierarchy has been established here and where breeders are located in it. One of the reasons is that property is more than legal text: the successful control of seed implies the collaboration of seed processors, farmers, plants and the state, not all of which are equally interested in extending and upholding breeders' rights. In order to work, rights need to be turned into functioning scripts, something the German seed sector struggled with in the aftermath of the 1991 UPOV reform. There are, of course, different ways of engineering scripts. If the economic and political landscape that seed circulates in cannot be remodelled, perhaps it is more effective to re-engineer the biology of seed itself. Ideally, hybrid seed is seed that will return to its breeder with or without the farmer's compliance – or rather, it will make the farmer return to buy new seed. In practice however, this biological re-propertization of seed, too, is subject to economic and political constraints: cost and price of seed, yield advantages, the differing financial resources among breeders, their fragile common interest.

Property and its redistribution, therefore, are also a practical issue. If breeders cannot tame and align the scripts woven into and around their seeds, their property will not work. From a political-economic point of view, such a conclusion could seem convenient: to achieve redistribution, we do not have to undo property because it will undo itself before long. What the crisis of PVP reveals, one could argue from this perspective, is first and foremost an exaggerated belief in property as a tool of power and capitalist appropriation (Latour, 1993, pp 173, 201). In framing property as a powerful machine that channels wealth from the weak to the strong, the political-economic critique takes it for granted, overlooking the fragility and increasing precariousness of its construction. Instead of despairing over property's power, it should instead acknowledge that this power will be undone before long. What is needed on the way there is not so much struggle as patience, aided by a sense of the weak spots of property that can readily be exploited.[43]

The Nagoya case resists such an all too convenient reply to classical political-economic concerns. If property is coming undone here, it is not through the disintegration of scripts that once coalesced to further the interest of wealthy multinationals and industrial nations. What we can observe here is rather that these scripts are themselves overwritten by their own logic. Rights in genetic resources, modelled after older forms of property, complicate and challenge the business models behind patents and PVP but reify the assumptions that underlie them: that a lack of compensation, redistribution and above all recognition can be addressed with the instrument of property. Neither could we argue that property, overtly or secretly, serves the interest of the powerful here. The problem is that propertization has led to an impasse, a lack of power to address the matter at heart. A mismatch of scripts and the inability to get them to work together frustrates the intentions behind the framework. The wish for recognition and compensation on the one hand and the practices and materiality of breeding on the other seem to be impossible to articulate with each other through property.

What connects the three examples of PVP revision, hybrid seed and Nagoya on a more fundamental level is that each of them is a showcase of the increasing inability of property to solve social problems. This inability is not limited to purely economic matters such as market failure and how to compensate for it. It also involves issues such as recognition, responsibility, control and identification, all of which property was thought to solve not too long ago. The disputes over resowing, the inability to re-engineer wheat in ways that make it compatible with established notions of private property and the breakdown of exchange in the face of Nagoya however suggest that it has become increasingly difficult for property to live up to such expectations. But if these are the symptoms of the 'property condition' of our times, what is the diagnosis and what are the causes?

Pottage (2006) and Perzanowski and Schultz (2018) rightly point out that the category of ownership has become tenuous. Yet this diagnosis does not go far enough: it only problematizes the subjective side of property relations. It is true that for various reason, the subjects of property relations are finding it more and more difficult to live up to their role as owners. Be it for overriding claims by others or for increasingly tenuous analogies between labour, invention, discovery, stewardship and creation, the owner appears as a fractured social figure. The same, however, might be said about the objects meant to be owned, which no longer seem to be able to comply with the assumptions of property thought. Where does property in seed end and where does it begin? On the farm? On the market? With a new generation of plants? In the faraway lands of its genetic ancestors? In the digitized sequence of its genome? These questions seem more difficult to answer today than 30 or 60 years ago.

But the problem does not end here. The crisis is not simply one of a deficit, of insufficient ownership or underperforming property objects. It is at the same time also one of a *surplus* of property: a continuing proliferation of property claims and forms; an increasing contradiction between previously demarcated rights and realms. There is not only too much ownership in theory (Pottage, 2006), there is also too much *property* in practice. If property has ceased to function properly, it is because there is an impasse between personal and intellectual property, between sovereign rights and the freedom to operate, between the right to exclude others from a patent and the privilege to appropriate competitors' varieties for one's own purposes. And yet the natural reaction of the plant breeding industry to this problem is to interpret it as a *lack* of property, one that has to be addressed with more, new, better, custom-tailored forms of property that will fill the gap between existing property regimes.

No matter that in practice we see the opposite result. That the strengthening of plant breeders' rights clashes with older, equally legitimate notions of property built around perfect alienation of goods and traditional resowing practices. That hybrid breeding does not only take away materially what UPOV 1991 takes away legally but might also consume breeders' companies and breeding pools in the process, not to mention the 'variety bargain' implicit in PVP law. That new property, rather than patching up old property, cannibalizes it. Still this cannibalism is an incomplete one. The new forms of property choke on the old ones, unable to fully swallow them yet unwilling to spit them out. The result is that this time it is not only the have-nots that find themselves on the losing end of property relations – they are increasingly joined by the haves. Unlike the new property of the mid-20th century, the property of today causes more problems for its owners than it can solve.

The true crisis of property can be seen in the fact that the failures of property are fuelling, not slowing, propertization. The fragility of property therefore cannot console us, since more is at stake than matters of distribution and, in any case, time will not solve the problem. What should concern us most is the increasing rift between property's exhaustion as a way to bring order to the world and its continuing appeal as an instrument of order. It is not that the people in the seed industry are not well aware of the limits of plant variety protection, patents or personal property. If there is no end to property's crises in sight, it is because despite all its shortcomings, they struggle to find a better answer to these crises than more property.

At the End of Property

The controversy comes to an end sooner than expected. On 29 June 2017, the EPO issues a statement (EPO, 2017), announcing that its Administrative Council has amended the office's implementing regulations (The Administrative Council of the European Patent Organisation, 2017). The new rule 28(2) of the regulations states that '[u]nder Article 53(b), European patents shall not be granted in respect of plants or animals exclusively obtained by means of an essentially biological process' (EPC 2000, r 28(2)). This step is a reaction to a clarification by the EU Commission from the previous year, in which the commission explained that the products of traditional plant breeding practices were not intended to fall within the scope of patentability when the Biotechnology Directive was drafted and passed (European Commission, 2016). As a first result, the EPO halted the proceedings of patents on conventionally bred plants in December 2016 (EPO, 2016); now it is making its U-turn perfect by completely overruling the *Broccoli/Tomatoes II* decision of its own Enlarged Board of Appeal from March 2015.

"My position in this matter is quite clear", Olaf Malek comments on the decision. "This in full contradiction to the Enlarged Board of Appeal, because of the Biotech Directive." He predicts there will be new complaints at the EPO seeking to clarify whether this rule is applicable in the first place: "Mr Kock has already requested in an appeals procedure that the Enlarged Board of Appeal will deal with the topic once more. So the last word is not spoken in the matter."[1] The EPO's turnaround is an attempt to revise the Enlarged Board of Appeal's decision, which has been heavily criticized from various sides, and to shortcut a lengthy legislative process in which the Biotechnology Directive and the European Patent Convention would have to be amended and clarified (Braun, 2023). As such, however, it is ill-devised. In strictly legal terms, the Administrative Council cannot overrule the Enlarged Board of Appeal as the supreme authority of patent matters, the EPO's implementing regulations or the EU Commission's opinion cannot overrule the Biotechnology Directive.

Nevertheless, the BDP officials are relieved over the EPO's move. "I would say that's a great success for us", Alexandra Bönsch says. She does not think that there is a contradiction between the *Broccoli/Tomatoes II* ruling and the regulation amendment: "We do not see a technical problem there because the Enlarged Board of Appeal is not precluded, meaning it decides anew in every case. And now it has this amendment as a further argument and can come to a different verdict in a new case." This saves the BDP and the seed industry from a strenuous battle to amend the biotech directive itself: "If it had come to opening up the Biotechnology Directive, we think support wouldn't have been that great. ... There would have been much more resistance from other industries and therefore we believe that this is a good solution for the time being."[2]

In the end, Bönsch will be right. Michael Kock, despite having left Syngenta before the EPO's change of mind, takes a refused bell pepper patent to the Technical Board of Appeal, which takes Olaf Malek's view in late 2018: the European Patent Convention and the Biotechnology Directive trump the patent office's self-imposed procedural rules.[3] The EPO's new president, António Campinos, in turn requests an authoritative interpretation from the Enlarged Board of Appeal. The board seizes the opportunity of the so-called *Pepper* case and declares the new rule valid in May 2020.[4] Applications submitted before the new rule will still be processed and granted, but from now on, the EPO will no longer issue patents on conventionally bred plants (the UK, which has left the EU in the meantime, and Mexico adopt the stance soon after). Applicants whose applications are still pending thus have no reason to challenge the rule, while everyone else has little incentive to sue in the first place. After more than two decades, the question raised by two inconspicuous vegetables is finally answered – for now.

It may still be the calm before the real storm. "We just kicked the can down the road", Michael Kock comments on the EPO's turnaround in 2018. He believes it is "a patch on a crumbling tube" that postpones the problem instead of solving it.[5] What Kock is hinting at is the coming of a new era of biotechnology, expected to revolutionize plant breeding once again. Over the last few years, so-called new plant breeding techniques have received increasing attention by scientists and the public (Lusser et al, 2012). Above all, gene editing with CRISPR/Cas and similar systems – methods that are able to target and modify specific loci and sequences in the genome – is hailed as a quantum leap in plant breeding (Hartung and Schiemann, 2014). Promising a quick, easy and cheap way of cutting and reassembling genes and genomes, it offers an alternative to the long and laborious procedure of introgressing and back-crossing traits found in exotic material. If the advocates of gene editing are right in their enthusiasm, Hubert Kempf would no longer have to go through breeding resistances into his germplasm over years; he could simply identify the gene and have it inserted into his high-yielding varieties in a 'cut and paste' fashion, a process that would cut trait development time in half. With R&D costs estimated around US$7 million per trait, this technology would be decidedly cheaper than old-style genetic modification yet much faster and more precise (Lassoued et al, 2019). For many seed companies, that makes it an attractive addition to the toolkit of conventional breeding.

If gene editing becomes an established technology in plant breeding, however, this will also prompt new challenges. While currently only about 1 per cent of all European plant varieties are affected by patent protection, Kock expects the share to rise to between 10 and 15 per cent with gene editing. This means that the effect of patents on plant breeding, currently felt only in a small number of crop species and sub-industries, will become much more manifest.[6] It also means that, unlike in conventionally bred plants, the legal discussions revolving around the meaning and scope of 'essentially biological' will not make a difference this time (Braun, 2023): the technical character of gene editing is beyond doubt, it will not be exempted from patentability within the frameworks of the EPC and the biotech directive. Patents on gene-edited plants, Kock tells me, will be "a storm against which discussions over native trait patents will feel like a mild breeze".[7]

"I've told [the breeders' associations], 'guys, this is a Pyrrhic victory. Maybe you got what you wanted. But you didn't get what you need'." Kock says. In his eyes, the EPO's turnaround is a ramshackle solution to an issue that demands a much more thorough approach. Breeders and their lobby have only focused on a single stick – the question of patentability – and closed the eyes to the bundle – all the rights that are tied to a patent that can interfere with the processes of plant breeding. With gene editing, there will be no easy way out of patents; the bundle can no longer be neglected in favour of the stick. Ultimately, Kock believes, the breeders have shot themselves in the foot: "There wasn't any proper problem analysis; no one asked: how do I solve this in a lasting way? And how do I solve it in a way that will still work in ten or 20 years?"[8]

The protest of plant breeders and activists has only put a halt to propertization momentarily. Yet Kock's remarks shed doubt on how long this breather will last. Under the EPO's new position, the possibility to appropriate traits with patents is tied to an element of technology in the breeding process. But this is currently the only obstacle in the way of patents – for genetically edited traits, the avenue to patent protection will be open again. The situation resembles that of the turn of the millennium, when health and environmental concerns were mobilized as an obstacle to patents in European plant breeding. But despite the scale of the anti-GMO movement's victory (ISAAA, 2018), it barely lasted a decade. This is why Kock would like European plant breeders and legislators to take an active instead of a defensive stance, to reform laws instead rather than hastily amend regulations. The whole bundle needs a makeover, not just a single stick. With immediate pressure taken off the sector and the EPO, such a legal overhaul is unlikely to occur anytime soon. However, the model of the ILP Vegetable has attracted attention from elsewhere in the industry: field crop breeders are trying to build a similar licensing platform for their seeds and patents. Yet some members of this initiative are trying to limit free use to the first generation of seed, anxious that they might hand over too much; to the disappointment of smaller companies in the club.[9] Meanwhile, the signatories to the CBD agreed to the extension of Nagoya to digital sequence information in late 2022 (Anderson, 2022), with unknown consequences for practice (Lawson and Rourke, 2016; Scholz et al, 2022).

In Germany, breeders have softened their stance toward the enforcement of PVP somewhat. "When I have visiting groups of farmers here and show them around and explain them the work that goes into making so many quality varieties, they're usually like 'Oh really, we weren't aware it was that much and that you were doing this much work; well, in that case ...'", Hubert Kempf says. Back when remuneration was introduced, he believes, "the stance was way too confrontational. It would have been necessary to explain more why – like it's increasingly done at the moment – how important breeding is, that farmers profit from breeding".[10] The dialogue with farmers, the BDP acknowledges, has indeed improved since the peak of the controversy over resowing fees.[11] Still, slowly but steadily, their claims have been reaffirmed by legislators, the courts and other actors in the seed value chain. Seed processors are now obliged to inform the seed trust about varieties, volumes and farmers upon legitimate request. What breeders have learned in the process, perhaps, is that property rights can bridge a lack of trust, but they cannot replace trust altogether.

Meanwhile, property is also shifting elsewhere. After the merger of Dow Chemical with Pioneer DuPont (now called Corteva) and the acquisition of Syngenta by state-owned ChemChina, Bayer has extended an offer to the shareholders of Monsanto to purchase the seed giant from St. Louis. After some deliberation with the market authorities, the Germans buy the Americans for US$63 billion, more that the global seed market's total volume at the time (Bonny, 2017). Christoph Then, the anti-patent activist, suspects that the prize Bayer is eyeing is not just Monsanto's patents, but, most of all, its biggest tangible asset: the company's unrivalled pool of germplasm. "If you put it in terms of products and valorization and resources and future, then Bayer is currently acquiring the world's largest seed bank in a sense."[12] But Bayer may well have acquired a giant liability instead. Ever since the takeover, the company has been mired in a series of lawsuits over Monsanto's Roundup herbicide and its possible links to cancer. As a consequence, the combined value of both firms has dropped below what Bayer paid for Monsanto (Elmore, 2021, pp 267–288). New patents are being filed, former competitors are being swallowed, the return of the patent question in plant breeding is already at the horizon. There is no end of property in sight. Or is there?

A crumbling Lockean topology

In this book, I have tried to highlight what property contributes to economic life – and beyond. As noted in the beginning the perspective is, of course, a partial one: that of seed, in Western Europe, primarily from the point of seed companies. It is also partial in its critique of property, questioning the notion of power that underlies many other critical and affirmative accounts, instead of offering a critical evaluation of property. More precisely, I have portrayed this lack of power as a *crisis*, characteristic of our times, rather than inherent in property across space and time. The scope and scale of this crisis only become apparent when juxtaposed with times and places where property worked or works better, or at least differently. Among other things, the aim of Chapter 3 was to provide a contrast between a time when seed could still be propertized without contradicting or overriding existing forms of property. If the failure of property is something that begs explanation, then we must also understand under what conditions property can work in the first place.

For the longest part of the modern era, property has been a key instrument for organizing, rearranging and formatting people and things. From the 18th century onwards, new forms of property were invented – copyrights, patents, trademarks – that allowed to appropriate ideas, forms and things in novel ways. As critical as we may today be of these forms, their effect was not only negative or disruptive. Authors' rights allowed artists to pass on something to their families and keep others from distorting creative works (Bellido, 2020); patents permitted new forms of economic activity and produced spillovers into an increasingly technologized society; trademarks ensured that buyers could make an informed choice (Cochoy, 2004). There is little doubt that without these new forms of property neither the modern film industry nor the pharma sector would exist in their current form. This does not mean that these property rights are unequivocally good or beyond debate. It is simply to note that they form a cornerstone not just of the modern economy but of a modern way of life, of 'a culture dominated by ideas about property ownership' (Strathern, 1988, p 18). Property is more than just rights or an auxiliary institution to the market: it is a very particular way of relating people to things, or subjects to objects (Strathern, 1988; 2022).

Plant variety protection, as I have portrayed it in Chapter 3, is a latecomer to this family of 'new property', emerging only in the 20th century. Although limited to a small market and its peculiar products (and therefore regarded as a curiosity by some), it has played an important role in this market and the modernization of agriculture in the second half of the 20th century (Harwood, 2012; Sanderson, 2017). PVP contributed (and continues to contribute) to a working market by turning it – historically, at least – into a *well-working* market. As a market object, seed encountered several problems: alienation, appropriation and exclusion did not work to the satisfaction of the market participants.

Between the established personal property of buyers and sellers in market goods, there was little space for another form of property that would override or outlive market exchange and its principles. In theory, the simultaneity of markets and persisting property ties is incompatible; seed can only fully belong to the buyer or the seller. And yet, PVP worked in practice, commodifying seed by de-commodifying it. The way PVP divided the bundle of rights attached to seed resonated with the division of practices between farming and plant breeding, which had developed during the professionalization of the latter. Within the socio-economic context of the time, PVP found a way for articulating particular modes of exclusion, alienation and appropriation in seed as a legal, commercial and biological object. The result was a property framework that, rather than disrupting them, mirrored the circumstances of its time.

The case of PVP may have been a rare historical exception for the 20th century, as the comparative history of US plant patents suggests. Often dismissed as an ill-institutionalized framework, its limited scope may perhaps rather point to the historical impossibility of building a commercial market and private industry for seed (Brandl, 2017, pp 109–131). Nonetheless, US plant patents were not simply a dead end. Early on, they acknowledged in law what PVP neglected, namely plant traits and their production. In the early 20th century, the obstacles to full patent protection of plants were considered too great; the compatibility of patents and plants was answered in the negative by US legislators. This was not to last: through the increasing technical transformation of plants, they moved closer to the realm of the patentable. At the same time, patents were preparing to meet plants halfway, moving from a specialized intellectual property right for inventors and inventions to a generalized instrument for turning any possible innovation into money. At the end of this development, even low-tech plants were deemed perfectly patentable, as most of the practical obstacles to patenting had either been resolved or pushed to the side by shifting the legal emphases.

This chain of translation ultimately led to the *Broccoli/Tomatoes II* case, in which the EPO approved European patents on native traits and conventionally bred plants. The intrusion of patents into the domain of traditional plant breeding poses a problem for the PVP framework and the practices it is meant facilitate, channel and support. As patent rights are much more restrictive than PVP, they threaten to interrupt the flow of germplasm and its collective improvement by plant breeding companies. Patents thus seem to be a prime example of 'anti-commons', an excess of private property that leads to an under-use of a good (Lametti, 2013; Aboukrat, 2021). A closer look at the value of patents, however, complicates the question of how to use them 'properly'. Companies, governments and research institutes have come up with very different ways of drawing value from patents. The problem with patents is not simply that they incentivize or inhibit the production and circulation of value. Depending on the context, one and the same patent can channel, signal or enclose vastly different values. Unless we stick to one particular perspective, this multiplicity makes it impossible to answer whether they do more bad or good. Since patents link (and at times short-circuit) incommensurable logics of valuation, however, it is also impossible to isolate one form of value and thus to reach a definite conclusion about what they are 'actually' for (Machlup, 1958; Biagioli, 2019).

Patents are a common scapegoat for many of the problems in contemporary plant breeding. Much of this criticism is justified, but to reserve it for patents would be wrong. Rather, I have argued in Chapter 6, the problem is a more fundamental one: not just of patents or intellectual property, but of property in general. Whether property takes the form of remuneration fees, hybrid varieties or Nagoya material, plant breeding increasingly shows signs of an over-saturation of property. Plant breeders encounter more and more difficulties to maintain and enforce property in their plants. With the Nagoya Protocol on biodiversity, they are suddenly confronted with property claims which resemble their own intellectual property, only that this time they find themselves at the opposite end of the property relation. The concerns and motives behind Nagoya are legitimate and rooted in a long history of skewed relationships between the countries that harbour the world's biodiversity hotspot and those that are home to industries that turn this biodiversity into money. Like in the cases of trait patenting or seed reuse by farmers, however, the issue cannot simply be addressed as one of legitimacy. Instead, it is the clash of equally legitimate concerns and interests that has plunged the seed sector into a dilemma. Caught between economic interests, the pride of owners and their desire to be acknowledged, there seems to be little space for reconciling all these different forms of property with each other. It is not only the future of plant breeding that is at stake here. In the process of making property in seed fit for the present, the industry also risks its past achievements.

Over the years and decades, it seems, property in plant breeding has turned from something that could solve social problems into something that instead creates more of them. This is certainly true in plant breeding, as could be seen in this book. But it can also be observed elsewhere: in the interference of digital copyright with our ability to use software, the internet or even just our purchases in the way we see fit (Perzanowski and Schultz, 2018); in disputes over the misappropriation and exclusivity of traditional knowledge and aesthetics (Rodriquez, 2006; Young, 2008); in the misalignments between patents, genes and our sense of property over our own body (Gold, 1996; Rabinow, 1997). Perhaps this impression is again owed to a partial perspective, which compares the controversies over property of today unfavourably with the settled property arrangements of the past. Yet this cannot explain why some of the property forms that were largely unproblematic in the past – such as patents and PVP – have become more controversial over time.

As Alain Pottage so aptly puts it, 'it is increasingly difficult to impose a Lockean topology on the world' (Pottage, 2004, p 3). What at least seemed possible or conceivable in the past – to add another or extend an existing form of property without drawing established ones into question – can no longer be achieved today. If the larger project of propertization entailed to turn everything on the globe into either owner or owned, its topology appears increasingly fragmented. Not just because there are more and more competing forms of property; there is also an increasing number of entities that fall in the cracks between subject and object, person and thing. Modern copyrights and patents, essentially developed against the backdrop of a late 19th- and early 20th-century industrial society and its property objects, seem to be unable to cope with the proliferation of things that cross the neat boundaries between consumers and producers, between using and reproducing, between knowledge and material. As a result, private property rights invade the private sphere of farmers, patents extend to the human body, sovereignty and property can no longer be squared (Aoki, 1996).

If the novel property of the 19th and 20th century was able to order and govern parts of economic life despite its contradiction to landed or tangible property, it was because there was still sufficient space for them to occupy. The gulf between the activities of industrial capitalists and Fordist consumers was large, or at least larger than that between providing and using software. The line between farming and breeding could be drawn and maintained along the farm fence because the revenues from selling seed stabilized this distinction instead of undermining it. Books were something that was much harder to reproduce than today; collecting and using exotic plant material required little consideration of others' claims. Today, these distinctions have shifted, blurred or collapsed. Cars and printed books, with their high entry costs for industrial production (Callon, 1994), are no longer the model for intellectual property to go by. Likewise, land and mobile possessions can no longer be taken as prototypes for a modern theory of property.

Instead of talking about cracks, we might as well consider what sprawls from underneath them. Over and over again, the bundle of rights and duties has been multiplied and divided. Since the introduction of plant patents and PVP in the mid-20th century, they have been joined by utility patents, protected geographical indications, the Nagoya Protocol, supplementary protection certificates (Correa and Correa, 2023), terminator genes (van Dooren, 2007b), chemical and cytoplasmic sterility systems and globally marketed trademarks such as Pink Lady; all of which could in theory coincide in the same plant. This does not just mean that plants are subject to a myriad of different, sometimes contradictory, property scripts. It also implies that whoever owns them also has to share them with someone else. As a result, both ownership and the objects of property become fragmented, while the bundle has turned into a veritable thicket. And yet this does not keep lobbyists, industries, well-meaning politicians, activists and, not least, scholars to ask for *more* property (Siegrist, 2006; Praduroux, 2017): property in our own body and its parts (Pollack Petchesky, 1995; Dickenson, 2017), ownership of data (Schwartz, 2003; Hummel et al, 2021), authorship of and control over customs and traditions (Young, 2008) or the use of blockchain technology to inscribe property into digital objects (Krasikov, 2022).

What has once again come to the fore over this process of fragmentation and sprawl is the social nature of property. The long battle that political economy has led against private property, be it as a concept or an institution of public life, may today seem strange in the face of things that do not fully belong to anyone. The entanglement of property has socialized owners and property in new, often unexpected and – above all – undesired ways. Bargaining for a license with patent holders, as we have seen in Chapter 5, can transcend even the capacities of moderately large seed companies. Whom a trademark for a particular franchise like Star Wars or Harry Potter belongs to depends on whether you want to print their characters on a T-shirt or put them in a box of cereals as well as on where you want to do it (Arnerstål, 2021). One of the results is that behind a seemingly innocuous toy for kids, there is usually an apparatus of corporate juggernauts who alone can navigate the thicket of international licensing. While this is still a concern for political economy, the larger problem this points to is that property is increasingly becoming too complex to handle.

And yet this does not silence the voices that demand even more property. We could dismiss such calls as the expressions of a 'property ideology' that are unable to think beyond ownership and possession. What I have tried to show in this book, however, is that the 'property condition' amounts to more than just an ideology. Intellectually indebted as we are to its idea of property and reliant as we are on its laws, practices and techniques in our everyday lives, we should better speak of a *culture* of property. Our collective inability to square our expectations of property with what it can realistically deliver, as well as our failing efforts to reinforce our crumbling property with more of it, cannot simply be denounced as an ideology. The crisis of property is a cultural crisis, one that generalizes Machlup's (1958) patent dilemma: differently but collectively, we all rely on property.

After property?

But is there a way out of the problems sketched in this book? If the crisis of property is indeed one of culture, of a way of life, then how can we address it? One obvious answer is that offered by the political-economic critique of property: where property does not live up to its promises, where it leads to abuse instead of use, it should be done away with (Loick, 2023). In the case of plants and seed, this poses the question of where we could replace property with something else that is more up to the task. We could discuss, for example, to entrust trait development to the state and its institutions again, rather than the market. As tax-funded entities, these public institutions would not need to rely on property to recoup their expenses. They could simply give away traits, perhaps also varieties, for free; no strings attached. Anyone could thus appropriate and use these public innovations without fear of ending up obliged or entangled. This is the solution that Jack Kloppenburg (2004) and Jonathan Harwood (2012) have pointed to for countering the trend toward private property rights in the seed sector.

The appeal of such a proposal is indeed that it does away with property altogether: if traits, varieties and genes are handed back to the state, the Gordian knot of property would at once be cut. Yet I am sceptical that a 'return to the state' can address the full extent of property's crisis. My first reservation is an empirical one: private capital has not simply seized seed from the state. The second half of the 20th century is rather characterized by an unwillingness of many states and public institutions to continue their mission of providing agriculture with 'free' seed. Olaf Malek, who impersonates the classic career of a patent attorney from public research into private counselling, does not believe that state funding will replace property any time soon. "I guess the Green Revolution in the 60s was essentially driven by such public institutions", he says. "But that's not the spirit of our times anymore."[13] Across industrialized countries, public research has retreated from applied plant breeding (Stengel et al, 2009; Harwood, 2012), and where it has continued its mission, it has not rejected but embraced patents and other forms of property (Heisey, King and Rubenstein, 2005). Neither the *Tomatoes* nor the *Broccoli* patent would exist without this embrace of property and its promises. Nagoya is another cautionary tale of the state's ability to sort out property.

This is, of course, merely a reservation, not a refutation: it is not to say that a return to 'truly' public research is impossible, just to note that in practice, the state is not necessarily the other to property and markets. Just like everyone else, public research is part of our property culture, not its antithesis. Imagining the state as the other of property does not only obscure the shared roots of the rule over people and the rule over things. All too often, it also disconnects the empirical reality of states from the theory of the state. If the state as a provider of goods essential for social life is supposed to be the antidote to property, it will have to be ingested in small doses, not swallowed whole.

Equally sceptical of private property and the state, some critical theorists are well aware of this. Their solution to the simultaneous excess and deficit of property is to resort to a different other, the commons. 'Commons differ from both private and public property because no one, neither a private individual nor a state institution, has the authority to deny use to others', Daniel Loick (2023) stresses. That there is no possibility to exclude others implies a different way of managing things and living together (Ostrom, 1990). The conflict and contradiction between interests, the figure of the commons implies, will not be answered in favour of only one side – the trait breeders, the variety producers, the farmers, the Global North or South – but resolved in the realm of a politics bound to practice: there is no commons without commoning (Helfrich and Bollier, 2021). The proposition of the commons is not to disentangle property by taking from the market and giving to the state; it is to entangle it even further by folding all the different subjects and objects of property into one. In the process, it is implied, both the interests of the subjects and the mutually overriding scripts of property can be aligned with each other.

One of the downsides of this 'collapse' of property in the commons is of course that it is ideally tied to a specific place and community: we are more likely to see it emerge in a particular, spatially limited setting that can still be governed with the deliberative tools that commoners have at their disposal. Nevertheless, there have recently been attempts to expand the logic of the commons beyond the local and to mobilize it against private property and its shortfalls. So-called open source seed initiatives have developed licenses for seed that are meant to be open to everyone (Kloppenburg, 2014; Kotschi and Horneburg, 2018). These licences, printed on the seed bags and considered effective upon opening them, take lessons both from so-called bag tags and creative commons licenses used for digital media. Like the licensing conditions printed on patented commercial seed bags by plant science multinationals, these scripts are not bound to a particular place but travel with the seed. Unlike these 'closed' licenses, open source seed licenses require users to offer their own seeds under the same conditions, thus steadily enlarging the pool of seeds and breeders in the sprawling commons. The legal basis for this, aside from contract law and terms of service, is the Nagoya Protocol, which the open source seed initiative interprets as giving them authority over the seed they have bred (Kotschi and Horneburg, 2018).

Importantly, these licenses contain a clause that prohibits the patenting (Kloppenburg, 2014), sometimes also the protection as a plant variety or a trademark (Kotschi and Horneburg, 2018) to prevent enclosure of and exclusion from seed. As a consequence, they also exclude a smaller or larger portion of commercial breeders from participating, depending on whether PVP is also prohibited under the license or not. This is a point of contention within the open source seed movement: should it be possible to demand royalties for one's own contribution through PVP, or is this idea already incompatible with open source seeds (Kloppenburg, 2014)? It is both a philosophical and a practical question, as the exclusion of PVP raises the question of funding for breeding programmes. The dispute highlights the downside of open source seed licences as an emancipatory and inclusionary idea. To allow the free use of seed, the terms of use have to be dictated; to prevent the exclusion from access to seed, certain actors need to be excluded. The commons in general and open source seed in particular are maybe best understood not as the antithesis to property but yet another form of it: neither one nor the other can make do without elements of inclusion and exclusion, of governing people through things and vice versa.

Although the primary criticism of political economy is aimed at matters of inequality and distribution, there is always an element of a 'freedom from property' that looms in the background: the freedom not to be exploited, not to be ruled, not to be restricted through property. The desires of those who reject and those who unanimously embrace property, it seems, are not that different. There is also, I believe, a shared conviction of the power of property; a conviction that is most apparent not in political-economic but in moral-ontological attempts to overcome – and at the same time reinforce – property. Discussions about 'rights of nature', that is, the status of non-humans as subjects with own legal standing, are sometimes difficult to entangle from arguments about the human communities they are embedded in. The key argument that has emerged in recent years, however, is that not only animals but also lakes, mountains, rivers or trees should indeed be recognized as sovereigns over themselves (Stone, 2010). Not only that: some authors even go so far as to argue that this should not just amount to subjective rights but to a simultaneous property title over themselves. Since nature would already own itself, it (and those who depend on it) could no longer be appropriated and exploited by someone else, the argument goes (Wesche, 2023) – an argument we have already encountered in the history of the Nagoya Protocol in Chapter 6. Again, it seems, property can only be made whole when subject and object are synonymized, when things become persons.

A while ago already, Marilyn Strathern (1988; 2022) has pointed out our difficulty of conceiving the world outside of this duality between property objects and subjects: they are the two modernizing frontiers we have at our disposal. To reject one is difficult without embracing the other. Yet this is not a strict dichotomy: to make ourselves recognized as subjects in our own right, we rely on the rhetorics and technologies of property. This is evident in the feminist movements that have to frame the female body as something that already belongs to its inhabitant and thus cannot be claimed by anyone else (Pollack Petchesky, 1995). It is also apparent in the long political struggle of women, freed slaves and other groups to complete their newly-won recognition as subjects by the ability to legally acquire property (Alexander, 1999). Again, this is not to reject or ridicule the arguments or struggles for recognition. It is simply to point out how strongly they rely on alternative forms and orders of property, rather than (as they often claim) true alternatives to property (Nichols, 2020). It is only through the juxtaposition of radical alternatives to property that its return through the backdoor emerges as a contradiction that requires legitimization. In trying to move beyond property by radical means, the accounts quickly find that they have returned to the point of departure.

Jack Kloppenburg (2014) wonders, citing the poet Audre Lorde, can the master's tools really undo the master's house? Looking at the fledgling open source seed movement, his answer is a cautious 'yes': if we do not understand property as monolithic, if we do not take at face value the claim that its ends are necessarily the interests of the powerful, if we re-appropriate, reassemble and redistribute select sticks in the bundle, we can address property's problems without ending property altogether. And yet he also acknowledges that others in his movement are less optimistic when it comes to finding an arrangement with PVP and commercial breeders. Many of the critics of property, it appears, have difficulty accepting that they live in a world also inhabited by property. Perhaps this yet another crisis of property: faced with the contradictions of property, the only alternative to multiplying property is eradicating it.

Conclusion

What Kloppenburg's pragmatism highlights, nonetheless, is that there might be different ways of processing the crisis of property in plant breeding as well as the other small and big crises of property elsewhere. Although on the face of it, we could think of these crises as matters of distribution or of recognition, they are in fact not only both (Loick, 2023) but more. What the sprawl of property has made obvious is that perfect property – that sole and despotic dominion – is a fiction that can only be upheld under rare circumstances. Every day, we instead discover that the things we own belong to someone else (Perzanowski and Schultz, 2018). The homes we thought ours turn out to be ruled over by banks and city councils; the state claims the inheritances our parents pass on to us; the varieties we have bred are subject to licence fees for the genes they contain. If there is no one to celebrate that the social nature of property has come to the surface, that we actually co-own everything, it is because of the horror of being at the mercy of others. If critics speak of a return of feudalism (Müller, 2006; Schubert,

Böschen and Gill, 2011), it is because we must fear that we, as persons, are again being ruled through things.[14]

What is at stake here, however, is not just material dependence and personal autonomy. Despite its central role in modern culture, property is not an end in itself. In the best case, it is a tool to make things work, to organize practice, to establish a flexible bond between subjects and objects. All too often, however, we take for granted that property will work, one way or another – even if the power of property is far from given. Property, or at least as I have portrayed it here, is a fragile and incomplete instrument, one that depends on a suitable context and the cooperation of a lot of different elements. What has driven propertization in the past is certainly not just greed, entitlement or ideology: it is also the experience that property can indeed work, even if it does not always work as intended. Who could blame chemical companies for extending their patenting activities to DNA and genes if patents made them good money in their chemical division? Who could criticize plant breeders for seeking to repeat the success of PVP? Who could not understand the desire of developing countries to establish a form of property as lucrative and prestigious as patents?

Yet in the process of trying to make property work *once more*, they forget what made it work in the first place. If patenting chemicals worked well, it was because they were far removed from the consumers who bought them. If PVP worked well in 1961, it was because there was still space between the nursery and the farm fence that no one had claimed yet. If patents on organic compounds and genes are lucrative and prestigious, it is because we all too readily forget about the patents that are devoid of any value. The less suitable the conditions for property as a tool for meaningfully governing things and people, the harder it is to make it work. And the harder we try to make it work, the greater the chance that we sacrifice the past achievements of property in the process.

Kloppenburg's and his allies' attempts to reassemble the bundle of scripts point the other way: not making property work by any means possible, but using the legal, practical and technical means at our disposal to make property work *better*. The same could be said about Michael Kock's project to slim down and streamline patents in plant breeding to prevent their collision with PVP by replacing the right to exclude with the right to be included. To make patents work for plant breeding, Kock is willing to give up on the biggest stick in the bundle. "I've always had my difficulties with the term 'property'", he admits. "That's why I always prefer 'intangible rights' or 'benefit sharing' or whatever you want to call it."[15] To make property work again, perhaps we will have to give up on the notion of property altogether: the word promises more than it can deliver. Understanding it as the unlikely alignment of a multitude of scripts that together can connect, divide and move people and things might give us a more precise idea of where the issue lies – and a more realistic picture of what property can and cannot do. Either way, we should be prepared to sacrifice property to save it.

Naturally, sorting out which scripts are and are not needed, where they should begin and end, and which ones we can prioritize over others without risking the failure of property will still be hard, frustrating and political work. Some scripts will have to go, while new ones will have to be assembled. Some scripts will refuse to work will others will run out of control. Over the process of propertization, both plant breeding and property have changed: it is difficult to imagine the former without the latter today. If we want to imagine the plant breeding of tomorrow, we will also have to come up with new ways of imagining and assembling property.

Notes

Chapter 1

[1] Interview November 2017, translated from German.

[2] Margaret Radin early on spoke of 'rhetorical gerrymandering' (Radin, 1993, p 14) in this context but also noted that it was not always successful, nor limited to economic interests and actors.

[3] Ultimately, therefore, this is also a critique that believes in economic rationality, only that it mistrusts the idea of (private) property as the instrument of such rationality (Posner, 1986; Landes and Posner, 2003) and puts it on its head.

[4] There are overlaps with the 'political-economic' tradition of critique in many of the texts from this tradition. Ultimately, however, the two disagree, as is evident especially in the case of surrogacy and markets for babies (Lewis, 2019).

[5] The present book owes a great deal to the ideas developed and discussed in the rather obscure field of 'cultural economy', a tradition developed largely in disagreement of 'political economy'. For an introduction, see du Gay and Pryke (2002) and Pratt (2009).

[6] Perzanowski and Schultz also cite economic reasons against exuberant intellectual property rights, such as poor people's access to educational resources or the economic benefits of second-hand markets to consumers and the economy. Overall, however, their misgivings are about the loss of a certain lifestyle, rather than the 'hard' deprivation political economy often focuses on.

[7] The list is of course incomplete and redundant, as Chapters 3, 4 and 6 will show. The difficulty of counting and delineating these different regimes is already part of the problem.

[8] To prevent the complexities and contradictions of property in seeds from emerging, already John Locke (2003) had to immediately have his fictional acorn eaten upon pick-up. Otherwise we might have started to wonder if the oak tree, too, is contained in it or whether picking up an acorn amounts to owning any future acorns it produces.

[9] This remains true if we attempt to become legal 'counter-experts' who seek to challenge the hegemonial doctrines in law: legal sociologists, legal anthropologists, philosophers of law or socio- and critical legal scholars. All these people still have to pass through the law in order to comment on property (Swedberg, 2003b, p 3).

[10] At the same time, the amalgamation of 'property' with 'rights' could leave us to wonder what, say, a sociology of property, as opposed to a sociology of property rights, could look like (Carruthers and Ariovich, 2004).

[11] Even accounts which claim to spell out the laws and principles of economics 'in their own terms' stop shy of applying the same rationale to property: compare, for example, the disagreement between new economic sociology (Carruthers and Ariovich 2004; Swedberg 2003a) and constructivist market sociology (Callon, 2021) with their agreement over property as a legal phenomenon.

[12] The very existence of lost-property offices is a striking testimony to the cultural value we place on property and its uncontestedness at least in certain realms of life. Conversely, archaeology, a lost-and-found office turned discipline, shows how the material practices of these places can inform a research programme (Klevnäs and Hedenstierna-Johnson, 2015; Earle, 2017).

[13] Anthropology (Hann, 1998a; von Benda-Beckmann et al, 2006a) and socio-ecological systems theory (Ostrom, 1990) have long stressed the important role of collective practices in property arrangements. What I argue in this book, however, is that there are aspects of property that are not captured by either practices or rules.

[14] The common denominator for most sociological analysis to this day is that property is an exclusive, and therefore asymmetric, legal relationship with regard to a good and resource (Carruthers and Ariovich, 2004). A proprietor's control over her property is legally sanctioned by the state, whose power monopoly ultimately lies at the heart of law here: 'The police apparatus would be mobilized, weapons inspected, warrants issued, jail keys turned' (Gouldner, 1970, p 307). I find that little can be gained from such a theory of property, at best a rather simplistic theory of state power (North, 1981). As we will see in Chapter 6, the problem starts with the fact that the police apparatus might never be mobilized.

[15] Readers familiar with STS will have little difficulty recognizing the hand of actor-network theory (Blok, Farías and Roberts, 2019) in this book, a school of thought that grew out of STS in the 1980s and 1990s into a broader alternative sociology (Latour, 2005), especially the strand that ventured into economics (Callon, 1998a; Barry and Slater 2005; Muniesa, 2015). While widely associated with a strong notion of the agency of things and a criticism of abstract categories like 'society', 'power' or 'the economy', it is perhaps best characterized by its annoying habit of claiming the opposite to whatever is received wisdom at the time of writing and passing this off as critique.

[16] The few exceptions to the former include Latour's discussions of forging informal intellectual property in molecular biology (Latour, 1984) and making hotel keys return to their owner (Latour, 1990), Schubert, Böschen and Gill's (2011) discussion of property as practice and maintenance as well as Slater's (2002) comments on the sales act as a passage of property. The most explicit and profound analysis of both intellectual and tangible property can be found in medical STS, however (Hoeyer, 2007; 2009; Parry, 2008). Intellectual property rights, especially patents, have overall featured much more prominently in STS (Bowker, 1992; de Laet, 2000; Hayden, 2003; Slaughter and Rhoades, 2004; Parthasarathy, 2017), but mostly as an 'exogenous' force that imposes itself on science. For a comprehensive overview, see Biagioli and Buning (2019).

[17] STS' discussion of law, to the extent that it exists, has largely followed ethnomethodologist traditions (Latour, 2010; Scheffer, 2010; Dupret, Lynch and Berard, 2015) or comparative and institutionalist political science (Jasanoff, 1997; 2007; Parthasarathy, 2017).

[18] Critical readers will note the relative absence of farmers' voices. Without doubt, this is one of the blind spots of this book. At the same time, however, it is deliberate: in giving room to the 'other' side of the seed business, I want to highlight its internal fault lines.

[19] This is also an assumption that lies at the heart of the distinction between belongings (property items of sentimental or use value) and property 'proper' like real estate, assets or wealth.

Chapter 2

[1] The processes of writing and discussing such texts in legal institutions are somewhat more difficult to observe and understand. For examples, see Latour (2010) and Scheffer (2010).

[2] A good example is the popular anthropological topos of gift economies and reciprocity (Strathern, 1988; Mauss, 2002): who could disentangle 'political', 'economic' or 'legal' logics here, let alone imagine a 'non-social' account of property rights?

[3] Time and again the assumption that the object and subject of property can be separated runs into difficulty, as Margaret Radin explains in her wonderful *Reinterpreting Property* (Radin, 1993).

[4] Although this is rarely made explicit, there is a widespread scepticism in the property-as-relation literature regarding the possibility of homogenizing all the complexities of social life in the first place. This melancholia is especially present in post-socialist accounts from Asia and Eastern Europe, which highlight the contradictions of both state and market property (Hann, 2006; Müller, 2007; Voicu and Vasile, 2022).

[5] The best example for this effect is probably Elinor Ostrom's (1990; 2005) work, which puts a much stronger emphasis on how the collective of owners should be governed than on what the materiality of fishing grounds or village pastures affords: little do we learn about the latter's ecology. Staying true to the property-as-relation tradition of thought, Ostrom's concern is with the collective of humans, not that of fish.

[6] Excerpt from a 2021 field lease contract; translated from German.

[7] It also allows the institute to plant wheat year after year without having to worry about crop rotation or the decrease of nutrients in the soil.

[8] If there is a general trend in propertization, it is certainly the increasing legal articulation of property through rights, whereas duties are either dropped or relegated to other parts of the law, such as consumer and environmental protection or work safety. The most vivid example and clearest formulation of this can be found in Coase's (1960) *Problem of Social Cost*, in which he dreams of property without liability.

[9] Another point of critique is the situatedness of the *bundle* image in Common Law, that is, US and British colonial jurisprudence. An attempt (perhaps simplified) at sorting out the legal traditions of property thought is to locate the *bundle* metaphor in Anglo-American law, whereas property as a person–thing relationship is rooted in Roman law. But there is also Germanic law, whose idea of property is more difficult to spell out in such terms (Siegrist, 2006, p 24). As Michele Graziadei (2017, p 81) points out, continental European legal thought does not mind (physically) splitting up property objects but takes issue with splitting the institution of ownership. Other legal realms might have even stronger objections to the idea of the *bundle*.

[10] Ostrom and Schlager still arrange the sticks in their property bundle into a hierarchy: the right to alienate is the right that is key to 'full' property (see Schlager and Ostrom, 1992, p 252). The relationship between alienation and 'property absolutism' will become clearer in Chapter 3.

[11] Although there are limits to the transferability of rights and duties, usually set by codified law, the freedom of contract as a legal principle in principle allows for an endless number of possible arrangements, subdivisions and transfers (contrary to the idea of a fixed number of property sticks, see Graziadei, 2017).

[12] This resonates with the different notions of performativity in social theory. See Dobeson, Brill and Braun (2023) on 'formatting'.

[13] On some of the problems with social constructivism, see Latour and Woolgar (1986, pp 273–285), Latour (1996) and Ian Hacking (1999).

[14] In their discussion of shifting intellectual property in digital things, Reitz, Sevignani and van den Ecker (2023) come very close to this conclusion – I only wish they would take the last step and speak of property (by other means), rather than 'functional equivalents'.

[15] The reason the problems raised by Latour's example have escaped property theory while Latour himself does not think of them as property problems is very likely that we usually

theorize property as revolving around exclusion of others and wilful alienation of our possessions, rather than locking ourselves out or losing our keys.

[16] Those annoyed by Latour's style and love for simplified, abstract examples can find a more complex study of property scripts and counter-scripts in Darla Thompson's (2014) *Circuits of Containment*, a fascinating if grim account of the technologies of antebellum US slavery and slaves' creativity in undoing them.

[17] Some readers will object here and draw a distinction between possession, which is only physical control, and 'actual' property, which is sanctioned by the law and thus 'legal' in nature. But as the saying has it, 'possession is nine-tenths of the law' (Singer, 2006): what the law adds, in most cases, is merely a stick to retrieve property after owner and possession have been separated.

[18] In the spirit of Pistor's (2019) *The Code of Capital*, I could also have used 'codes' instead of 'scripts' here. If I have opted for the latter, it is mainly because of its ambiguous meaning as well as a scholarly tradition that emphasizes how scripts are above people's intention, not the other way around: not everything can be written into a script and as we will see, some scripts cannot be mastered. This is at odds with Pistor's idea of 'code masters', who can more or less change property rules and relations at will (Pistor, 2019, pp 158–182).

[19] The bundle of scripts thus reproduces the problem already encountered with the bundle of rights and duties and that of practices: what could *not* be described as such a bundle? This is a challenge especially when we are dealing with relations that look a lot like property at first sight but should not be confused with it, such as marriage, stewardship or tutelage.

Chapter 3

[1] Least-square means are a method for calculating mean values of a series of measurements that deviate from a fitted curve. Since the results are sometimes clearly different from the arithmetic means, some breeders use this difference as an argument when pleading for market admission with the Federal Variety Office.

[2] Others contest the distinction between breeding and agriculture, and with good reason (van Dooren, 2007a, pp 23–26). As will become apparent in Chapter 6, plant breeding and agriculture cannot be fully disentangled.

[3] It should not be left unsaid that over the decades of his career, Hubert's wife has equally sacrificed her weekends; either by taking care of their children or by travelling with him to evaluate experimental wheats.

[4] For other agricultural crops like maize, sugar beet or rape, things look quite different: here the seed market is split between only a handful of companies. If wheat is exemplary for the diversity and competitiveness of European plant breeding (Ragonnaud, 2013), maize illustrates the opposite: a strong tendency toward market concentration, economies of scale and internationalization (Ipsen, 2016; Brandl, 2017; Brandl and Glenna, 2017). There is a considerable number of formidable works detailing the history and peculiarities of select crops and national seed sectors. For maize, see among others Fitzgerald (1990), Ipsen (2016), and Torma (2017; 2018); for wheat, Timmermann (2009), Head, Atchison and Gates (2012) and Müller (2013). Waltenberger (2020) offers a history of German rapeseed, Saraiva (2016) one of German potatoes; for Dutch hybrid potatoes, see the work of Stemerding and colleagues (2023). Chapman (2018; 2022) offers ethnographic accounts of Gambian breeding and PVP; Stone and Glover (2017) one of rice in the Philippines. Bonneuil and Thomas detail the history of the French seed industry in two volumes (2009; 2012), as do Wieland (2004; 2006; 2009; 2011) and Harwood (2012) for Germany.

[5] Castration and deliberate cross-fertilization of plants is simpler in some crop species and more difficult in others. The male flowers of maize, a dioecious plant, are easily removed, for example, whereas grasses pose an even greater challenge than wheat due to their even smaller flowers.

[6] A few basic predictions can of course be made. For example, cards that are very close to each other in the original stack are relatively likely to end up together. The hereditary behaviour of a few well-known and genomically mapped alleles can also be estimated statistically, though never individually. Due to the sheer number of chromosomes and genes involved, the outcome of an experimental cross is much more easily tested than it is calculated.

[7] Gene editing, that is, advanced molecular tools for transferring single genes between genomes, promise to be much more reliable and at the same time cheaper than classical genetic engineering. Industry experts estimate that research and development amount to US$7.5–10.5 million with gene editing, spread over a development time of about three years (Lassoued et al, 2019, p 49). See Chapter 7 for the implications of these technologies for property in plant breeding.

[8] Commercial plant breeding, at least in medium-sized firms like Hubert's, has until recently made limited use of genetics as scientific knowledge. As a university-trained breeder with a PhD, Hubert is of course familiar with genetics, but there is clearly a divide between fundamental research and applied breeding in the seed industry. It is often faster and cheaper to try a particular cross in the field than to develop a method for predicting it in the lab. In recent years, however, progress in marker-assisted breeding (Reece and Haribabu, 2007) has led to decreasing costs and more widespread application of molecular methods that help track individual traits in the genome long before they become visible in the field.

[9] Again, methods differ between breeders and crop species: some breeding programmes will follow the so-called bulk method of multiplying offspring over three generations before starting to select them; the more rigorous pedigree method will discard individual plants early on (Becker, 2019, p 269). Plants like rye, potato or maize, which have a different reproductive system from wheat, require more complex arrangements but follow the same principle of increasing diversity before selecting for overperformers (Acquaah, 2012, pp 337–381).

[10] See Becker (2019, p 269) and Acquaah (2012, p 311) for schematic illustrations of the breeding process in self-pollinators.

[11] Cultivated plants often hold on to their seeds for longer than their wild relatives: the shattering of ears, which benefits the dispersal of seeds in a cereal's habitat, was one of the first traits to be removed from cultivars through human selection (Becker, 2019, p 10).

[12] According to Bonneuil and Thomas (2012, p 18), the United Kingdom was ahead of Germany when it came to the adoption of commercial seed, whereas France kept lagging behind until World War II.

[13] Although West Germany was the first European country with a fully-fledged seed marketing law, other countries had anticipated similar solutions earlier. As a very early precursor, a papal edict from 1833 granting rights for (among others) agricultural discoveries is often cited (Heitz, 1991, p 19). Commercial breeders had advocated for dedicated IP rights early on through associations such as ASSINSEL (Sanderson, 2017, p 33); a German law including breeders' rights was drafted in 1930 but could not be passed before the National Socialists rose to power (Metzger and Zech, 2016, p 8). In the late 1940s, French breeders were pursuing a corporatist solution through the Caisse de gestion de licences végétales (Bonneuil and Thomas, 2009, p 114).

14 Plant breeders master this challenge by running a translocal testing apparatus of their own, seeking to anticipate and double-check the performance of their plants in the national tests.

15 Here we find commodification in its original meaning of French *commodité* (convenience). Commodities are a class of goods that are easily appropriated and can do with as little contextualization work as possible (Callon and Muniesa, 2005). For another example, see Verriet's (2013) work on how the convenience meals commodified cooking in the decades after World War II.

16 Schriftlicher Bericht des Ausschusses für Ernährung, Landwirtschaft und Forsten (19. Ausschuß) über den Entwurf eines Gesetzes über Sortenschutz und Saatgut von Kulturpflanzen (Saatgutgesetz). Bundestag Drucksache Nr 01/4339, p 5.

17 Schriftlicher Bericht des Ausschusses für Ernährung, Landwirtschaft und Forsten (19. Ausschuß) über den Entwurf eines Gesetzes über Sortenschutz und Saatgut von Kulturpflanzen (Saatgutgesetz). Bundestag Drucksache Nr 01/4339, p 5.

18 268. Sitzung des Deutschen Bundestages, p 13206. This is a rare occasion in the world of property in which the fence became a symbol of unity and continuity, rather than exclusion and separation. Compare also Blomley (2013).

19 Between the passing of the Seed Act in 1953 and the late 1980s, the Bundestag rarely discussed PVP, usually when a new revision of the UPOV agreement needed to be implemented (which happened in 1968 and 1985 and mostly concerned technicalities such as the extension of the term of protection). No attempts were made by any of the parties to renegotiate the bargain between farmers and breeders from the 1950s at these occasions. See https://dip.bundestag.de/erweiterte-suche?term=Sortenschutz&start=1&rows=150&sort=basisdatum_auf

20 My definition of a quasi-commodity differs from that of Schaniel and Neale (1999) in that I argue that quasi-commodities will in fact behave like 'proper' commodities: in practice, the two are indistinguishable. It is only when practice changes and runs into dead-ends and contradictions that the true nature of quasi-commodities comes to light.

21 Neither, however, is intellectual property simply ownership of a 'master artefact', as Peukert (2021, pp 52–56) postulates. The distinction between master and secondary artefacts is a tenuous one. In practice, the only difference is the amount of work and investments required to go from the latter to the former, as reverse-engineering R&D departments well know (Callon, 1994).

22 An average estimate for the industry puts the breeder's license at 19 per cent of the seed price at point of sale (DLZ, 2010).

23 The propertization of things for the purpose of commercializing them is often likened to the process of enclosing a piece of land (Landes and Posner, 2003, p 16; Mercier, 2021), echoing an old debate about the benefits of private ownership in landed property (Aboukrat, 2021). To render the image more complete, what is tacitly implied is that somewhere in the fence there is an entrance with a kiosk where visitors can pay the appropriate amount of money to enter the property.

24 As Sanderson (2017, pp 81–112) shows, the preconceptions of UPOV's PVP framework cannot always easily be translated to economic contexts different from the European one implicitly assumed by PVP – it can be questioned whether the global expansion of UPOV was always in the interest of developing countries.

Chapter 4

1 Attentive readers will feel reminded of Latour's (2005, p 63) dictum that 'action is overtaken by other agencies'.

2 'Variety', 30 years before UPOV, obviously was closer to the original meaning of something that deviated from an object of the same class than to the quasi-commodified seed in mid-century Western Europe.

[3] The historical contingency of today's separation of patent and copyright law is still apparent here. See Sherman and Bently (2003).

[4] One of the precursors to intellectual property is the physical ownership and monopoly over technologies that serve as links between form and exemplar, such as printing presses (Sherman and Bently, 2003). See also Peukert (2021), who bases his ontology of intellectual property law on this original hierarchy between form and exemplar.

[5] Neither Burbank's vocal opposition against plant patents nor his (by the time of the passing of the act) loss of standing in scientific circles did much to prevent him from becoming the 'model breeder' for this piece of legislation (Fowler, 2000, pp 637f).

[6] One of the most telling examples of failed fencing is the case of Stark Brothers, a major US fruit breeding company. In 1930, it had purchased an apple tree sprung from a bud mutation for 6,000 USD and put it behind a tall fence to prevent others from taking grafting wood; with little success (Fowler, 2000, pp 629).

[7] Burbank, for example, was familiar with Mendelism but also believed in a Lamarckian retention of acquired traits (Fowler, 2000, p 638).

[8] The split between plant breeding and plant genetics is pronounced to this day. In spite of all rhetoric of plant breeding as 'applied plant genetics' (Sánchez-Monge, 1993, p 4), breeders and molecular biologists draw from very different knowledge reservoirs and follow different practices for manifesting their 'shared' object, the plant. In many crop species, contact between both communities is sparse and selective; plant breeders frequently have an outdated and very theoretical knowledge of plant genomes while geneticists often lack understanding for the necessities and practicalities of working with phenotypes. Although genomics, genetics and applied breeding have moved towards closer collaboration in recent years, their division of labour testifies to their differences. See Timmermann (2009) and Lurquin (2001) for accounts of the respective knowledges and practices.

[9] To this day, even accounts like Fox Keller's (2002) or Dupré's (2004) have to start out from a simple, reductionist notion of the gene to subsequently complicate it. One could blame the prevalence of simplified popular imaginaries of genetics, but I would rather argue that it is the power of concepts themselves in whose face we cannot help but follow the path they pave. See Stengers (2011, pp 19–41).

[10] Early Belgian experiments (1950s to 1970s) initially produced promising results, which today nevertheless are regarded as artefacts, as they involved whole plant cells, rather than 'unfenced' protoplasts (Lurquin, 2001, pp 8–24).

[11] This also required the removal of the oncogenes from the Ti plasmid, a highly difficult technical challenge that Monsato beat their competitors to.

[12] In turn, plant breeding companies like Pioneer Hi-Bred and DeKalb also started to invest in molecular biology (Kloppenburg, 2004, p 217).

[13] This unanswered question would reemerge a few years later in the *Harvard OncoMouse* case at the Canadian Supreme Court (Kevles, 2002; Jasanoff, 2007, pp 211–213).

[14] For a detailed overview of the EPC's history and political-legal architecture, see Parthasarathy (2017).

[15] In effect, this encouraged patent applications which sought to claim a broad range of varieties and plant species as their field of application. The amended *Broccoli* patent, for example, does not only claim broccoli with higher glucosinolate levels, but also all other *Brassica oleracea* subspecies (including cauliflower and cabbage) and indeed any Brassica plant with such characteristics (G 0002/13, pp 12f), such as rapeseed and rutabaga (*B. napus*), turnip (*B. rapa*) or mustard (*B. nigra* and *B. juncea*).

[16] Interview with plant breeding company, December 2017.

[17] Unlike in later times, the issue was also not given much importance by European patent law in the 1960s. Interview with patent lawyer, June 2018.

Chapter 5

[1] Interview June 2018, translated from German.

[2] Interview November 2017.

[3] Interview June 2018, translated from German.

[4] Interview June 2018, translated from German.

[5] Interview September 2017, translated from German.

[6] Interview September 2017, translated from German.

[7] Interview March 2015. This practice is especially widespread in self-pollinating cereals like wheat and barley and takes place in the last two years of variety development when a variety's properties can already be assessed with high certainty.

[8] Interview September 2017, translated from German.

[9] Many anthropologists will no doubt shiver over this comparison, which of course is overly simplified for the purpose of illustration. To a large part, the gift economy relies on implicitness and the 'misrecognition' (Bourdieu, 2008, p 105) of reciprocity, whereas breeders here are neither unaware nor ashamed of expecting something in return. Then again, the concept of the gift as a human universal is not unproblematic in the first place (Strathern, 1988, pp 3–21; Thomas, 1991, pp 7–34).

[10] Patents on conventionally bred plants can be retrieved from PATSTAT (2018 Spring data; de Rassenfosse et al, 2014) by asking for all plant breeding patents (category A01H and subcategories) whose cooperative patent classification does not include biotechnology (C12N and subcategories), symbiosis with bacteria (A01H 3) or tissue culture (A01H 4). To further specify the search, I excluded algae (A01H 13) and fungi (A01H 15). Patents referring to processed plant materials instead of germplasm (F and subcategories) were excluded as well as patents for processes and products involving marker-assisted breeding (C12Q).

[11] Head of IP, interview December 2017.

[12] Of course the same can be said about the complementary narrative put forward by the plant breeding lobby.

[13] Interview October 2015, translated from German.

[14] Interview December 2017.

[15] Interview September 2017, translated from German.

[16] Interview August 2017, translated from German.

[17] Interview August 2017, translated from German.

[18] Field notes May 2017, paraphrased and translated from German.

[19] Field notes May 2017, paraphrased and translated from German.

[20] Field notes May 2017, paraphrased and translated from German.

[21] The cited breeder regrets the takeover mainly because of the change in corporate culture: "Syngenta never calculated", he says, "but the Chinese know very well how to calculate!" The first victim, he adds, is Syngenta's generous company car policy. Field notes May 2017, paraphrased and translated from German.

[22] Interview August 2017.

[23] Interview June 2018, translated from German. In a later mail (May 2020), Kock adds a qualification: no *educated* investor will accept a mere quantity of patents as a performance indicator.

[24] Interview June 2018, translated from German.

[25] Interview June 2018, translated from German.

[26] PATSTAT Spring 2018 data, global patent applications and grants for Syngenta. The company's applications drop from 1430 in 2008 and 1077 in 2009 to 399 in 2016 and 116 in 2017. Grants, naturally lagging behind applications, drop from a peak of 701 in 2008 to a low of 71 in 2016.

[27] Interview June 2018.

[28] Interview September 2017.

[29] Interview June 2018.

[30] Interview August 2017.

[31] Interview September 2017, translated from German. At the time of the interview, Monsanto was notably absent from the ILP Vegetable membership – as is its parent company Bayer Crop Science at the time of writing.

[32] Interview June 2018, translated from German.

Chapter 6

[1] Before and after the ruling in *Broccoli/Tomatoes II*, European and national plant breeders' associations lobbied Brussels and their respective parliaments to preclude the products of conventional breeding from patentability and to introduce exemptions and compulsory licenses that would allow the use of patented traits.

[2] This understanding of farming and the rights it entailed was not limited to the UPOV countries. Court rulings under the US Plant Variety Protection Act of 1970, which took the UPOV framework as its model, regularly upheld the right of farmers to reproduce, save and resow purchased seed (Sanderson, 2017, pp 235f).

[3] To get an overview of the debate at the time, I conducted a search in the wiso database for German print media outlets (http://wiso-net.de, 8 August 2018) on the keywords 'UPOV' and 'Sortenschutz' (plant variety protection). Early accounts, as far as they are critical and in the popular press, can only be found in *die tageszeitung* and seem to be largely informed by observations of Genetic Resources Action International (GRAIN), an agricultural activist network founded in the early 1980s (Peschard and Randeria, 2020). Except for Leskien (1991), however, they do not comment on the changes in the UPOV rules themselves but focus on the threat for PVP from patent law instead. Remarkably enough, I could not even find any articles in German agrarian magazines on the topic up to 1992.

[4] Interview April 2015, translated from German.

[5] Interview April 2015, translated from German.

[6] Interview April 2015.

[7] Interview September 2017, translated from German.

[8] The discussion probably emerged out of the discussions around the Biotechnology (rather than the CPVR) Directive and the contradictions of patent and PVP laws with regard to resowing. It is also possible that the farmers' privilege was excluded from UPOV 1991 to prevent a potential conflict between free resowing in PVP and exclusive use in patent law to emerge. The question of patent exhaustion in resowing was only answered very late, however, in *Bowman v. Monsanto Co.* (Peavey, 2014; Perzanowski and Schultz, 2018, pp 163f).

[9] On the difficulty of even accessing data on land ownership in Germany, see Sommer (2022).

[10] Although perhaps the most prevalent criticism of patents on seeds, the idea that patents could be used as an instrument of control over consumers was absent from (and certainly alien to) early anti-patent thought and movements as summarized by Parthasarathy (2017, pp 22–25).

[11] Interview April 2015.

[12] Interview April 2015.

[13] Interview April 2015, translated from German.

[14] Interview April 2015.

[15] Interview September 2017, translated from German.

[16] Interview 2015, translated from German.

[17] Interview September 2017, translated from German.

[18] Field notes July 2016, translated from German.

[19] On recognition in plant breeding, see Gill et al (2012) and Chapman (2018).

[20] Interview September 2017.

[21] Hybrid breeding is not the only example of property protection by biological means. In the case of plants, so called 'terminator' or genetic use restriction technology (GURT), which can prevent plants from reproducing altogether, has historically made headlines (Kloppenburg 2004, pp 319f; van Dooren, 2007b; Pottage 2011) and is often lumped with hybrid breeding in popular imagination.

[22] These genes need to be bred into so-called restorer lines, which are used in seed multiplication to produce commercial grain. To reproduce the male-sterile mother line, another set of plants, so called 'maintainers' are required, which do not contain a restorer gene but have a fertile cytoplasm. Cross-fertilization with these maintainer lines will reproduce the mother line true to form. See Tsunewaki (2003) for a detailed technical explanation of hybrid wheat systems.

[23] Field notes August 2018.

[24] Interview September 2017, translated from German.

[25] Field notes March 2015.

[26] Field notes March 2015, translated from German.

[27] Field notes July 2016, translated from German.

[28] Interview June 2018, translated from German.

[29] The discourse over biopiracy often implies that companies primarily engage in bioprospecting expeditions that venture out into biodiversity hotspots to collect, screen and ship plants and animals like colonial naturalists did in previous centuries (Schiebinger, 2004, pp 35–44; Pannhorst, 2016). Many genetic resources are not retrieved from their countries of origin as part of an explicitly commercial project, however. Plant science companies often obtain plants from seed banks or the personal collections of university scientists, which serve as intermediaries. The processes of collection and commercialization are thus decoupled spatially and temporally, some current plant products go back to collections made decades before the 1992 Rio Summit (Leberecht and Meienberg, 2014; Braun, 2021b, p 71).

[30] Dutfield (2003, p 3) mentions that in some cases, landraces and crop varieties allegedly even received variety protection without undergoing further breeding in their countries of origin.

[31] This is of course a simplification of Coase's property framework. It is perhaps more aptly characterized as a rigorous critique of fellow economists who prematurely call for liabilities or state services where the market seems to fail. Coase (1937; 1974) instead argues that in most textbook cases, such solutions are contingent rather than inevitable.

[32] Interview September 2017, translated from German.

[33] Interview September 2017.

[34] While a similar accumulation effect can occur with patents (for example through breeding several patented traits into one and the same plant), it is limited by the patent term of 20 years. In practice, only a handful of patents will thus affect a particular plant material.

[35] Interview December 2017.

[36] Interview June 2018, translated from German.

[37] Interview June 2018, translated from German.

[38] Interview December 2017.

[39] Interview September 2017.

[40] The USA and Canada have not signed the Nagoya Protocol as of December 2023.

[41] Interview September 2017.

[42] Interview June 2018, translated from German.

[43] Such patience is of course the patience of those who can afford to wait and placate. The great strength and at the same time weakness of Actor-Network Theory's concept of

power is that it is conceptualized from the point of view of the powerful (Latour, 1993, pp 208–209; Star 1990).

Chapter 7

[1] Interview November 2017, translated from German.
[2] Interview September 2017, translated from German.
[3] T 1063/18 (Extreme dark green, blocky peppers/SYNGENTA).
[4] G 3/19 (EPO 2020).
[5] Interview June 2018, translated from German.
[6] Interview June 2018.
[7] E-mail August 2023.
[8] Interview June 2018.
[9] Interview February 2022.
[10] Interview April 2015, translated from German.
[11] Interview September 2017.
[12] Interview August 2017.
[13] Interview November 2017, translated from German.
[14] I owe this formulation to Doris Schweitzer.
[15] Interview June 2018.

References

Aboukrat, A. (2021) 'Tragédie des anti-commons', in M. Cornu, F. Orsi and J. Rochfeld (eds) *Dictionnaire des biens communs*, 2nd ed, Paris: puf, pp 1274–1278.

Acquaah, G. (2012) *Principles of Plant Genetics and Breeding*, 2nd ed, Hoboken, NJ: Wiley.

Aistara, G.A. (2012) 'Privately public seeds: Competing visions of property, personhood, and democracy in Costa Rica's entry into CAFTA and the Union for Plant Variety Protection (UPOV)', *Journal of Political Ecology*, 19(1): 127–144.

Aistara, G.A. (2014) 'Latvia's tomato rebellion: nested environmental justice and returning eco-sociality in the post-socialist EU countryside', *Journal of Baltic Studies*, 45(1): 105–130.

Akkermans, B. (2017) 'The numerus clausus of property rights', in M. Graziadei and L.D. Smith (eds) *Comparative Property Law: Global Perspectives*. Cheltenham: Edward Elgar Publishing, pp 100–120.

Akrich, M. (1992) 'The de-scription of technical objects', in W.E. Bijker and J. Law (eds) *Shaping Technology/Building Society: Studies in Sociotechnical Change*, Cambridge, MA: MIT Press, pp 205–224.

Alexander, G.S. (1999) *Commodity and Propriety: Competing Visions of Property in American Legal Thought, 1776–1970*, Chicago, IL: University of Chicago Press.

Anderson, S. (2022) 'Stevia as a genetic resource: Intellectual property and Guaraní strategies for access-and-benefit sharing in Paraguay and Brazil', *Environment & Society Portal*, available from: https://www.environmentand society.org/arcadia/stevia-genetic-resource-intellectual-property-and-guar ani-strategies-access-and-benefit [Accessed: 3 December 2023].

Aoki, K. (1996) '(Intellectual) property and sovereignty: Notes toward a cultural geography of authorship', *Stanford Law Review*, 48(5): 1293–1355.

Aoki, K. (2008) *Seed Wars: Controversies and Cases on Plant Genetic Resources and Intellectual Property*, Durham, NC: Carolina Academic Press.

Appadurai, A. (1986) 'Introduction: Commodities and the politics of value', in A. Appadurai (ed) *The Social Life of Things. Commodities in Cultural Perspective*, Cambridge: Cambridge University Press, pp 3–63.

Arnerstål, S. (2021) *International Trademark Licensing*, Alphen aan den Rijn: Kluwer Law International.

Aubertin, C. and Filoche, G. (2011) 'The Nagoya Protocol on the use of genetic resources: One embodiment of an endless discussion', *Sustentabilidade Em Debate*, 2(1): 51–63.

Barry, A. and Slater, D. (eds) (2005) *The Technological Economy*, London: Routledge.

BDP (2015) *BDP-Position zur Ausgestaltung des Patentschutzes in der Pflanzenzüchtung: BDP fordert Beschränkung des Patentschutzes auf technische Erfindungen*, Bonn: BDP.

BDP (2016) *Tag der Biologischen Vielfalt – Ohne Zugang zu genetischen Ressourcen bleibt der Fortschritt in der Pflanzenzüchtung auf der Strecke*, Bonn: BDP.

Beadle, G.W. and Tatum, E.L. (1941) 'Genetic control of biochemical reactions in neurospora', *Proceedings of the National Academy of Sciences*, 27(11): 499–506.

Beck, P. (2016) 'Restopia: Self-storage as urban practice: "Like a hotel – but for things"', in C. Lewe, T. Othold and N. Oxen (eds) *Müll: Interdisziplinäre Perspektiven auf das Übrig-Gebliebene*, Bielefeld: Transcript Verlag, pp 117–140.

Becker, H. (2019) *Pflanzenzüchtung*, 3rd ed, Stuttgart: Ulmer.

Beckert, J. (2008) *Inherited Wealth*, Princeton, NJ: Princeton University Press.

Bellido, J. (2020) 'Experimenting with law: Brecht on copyright', *Law and Critique*, 31(2): 127–143.

Bellido, J. and Sherman, B. (eds) (2023) *Intellectual Property and the Design of Nature*, Oxford: Oxford University Press.

Bent, S.A., Schwaab, R.L., Conlin, D.G. and Jeffery, D.D. (1987) *Intellectual Property Rights in Biotechnology Worldwide*, Basingstoke: Palgrave Macmillan.

Bently, L. (2013) 'Trade secrets', in H. Howe and J. Griffiths (eds) *Concepts of Property in Intellectual Property Law*. Cambridge: Cambridge University Press, pp 60–93.

Bently, L. and Sherman, B. (2014) *Intellectual Property Law*, 4th ed, Oxford: Oxford University Press.

Bently, L., Sherman, B., Gangjee, D. and Johnson, P. (2018) *Intellectual Property Law*, 5th ed, Oxford: Oxford University Press.

Bently, L., Sherman, B., Gangjee, D. and Johnson, P. (2022) *Intellectual Property Law*, 6th ed, Oxford: Oxford University Press.

Berlan, J.-P. and Lewontin, R. (1983) 'Hybrid corn revisited', unpublished paper.

Bernhardt, S. (2005) 'High plains drifting: Wind-blown seeds and the intellectual property implications of the GMO revolution', *Northwestern Journal of Technology and Intellectual Property*, 4(1): 1–13.

Berry, D.J. (2023) 'Scientific activism and intellectual property: How UK-based agroecologists and plant synthetic biologists have challenged the status quo', in J. Bellido and B. Sherman (eds) *Intellectual Property and the Design of Nature*, Oxford: Oxford University Press, pp 289–315.

Bertacchini, E.E. (2008) 'Coase, Pigou and the potato: Whither farmers' rights?', *Ecological Economics*, 68(1–2): 183–193.

Bhandar, B. (2018) *Colonial Lives of Property: Law, Land, and Racial Regimes of Ownership*. Durham: Duke University Press.

Biagioli, M. (ed) (1999) *The Science Studies Reader*, New York: Routledge.

Biagioli, M. (2006) 'Patent republic: Representing inventions, constructing rights and authors', *Social Research: An International Quarterly*, 73(4): 1129–1172.

Biagioli, M. (2015) 'Patent specification and political representation: How patents became rights', in M. Biagioli, P. Jaszi and M. Woodmansee (eds) *Making and Unmaking Intellectual Property*. Chicago: University of Chicago Press, pp 25–40. https://doi.org/10.7208/9780226172491-003

Biagioli, M. (2019) 'Weighing intellectual property: Can we balance the social costs and benefits of patenting?', *History of Science*, 57(1): 140–163.

Biagioli, M. and Buning, M. (2019) 'Technologies of the law/ law as a technology', *History of Science*, 57(1): 3–17. https://doi.org/10.1177/0073275318816163

Biagioli, M., Jaszi, P. and Woodmansee, M. (eds) (2011) *Making and Unmaking Intellectual Property: Creative Production in Legal and Cultural Perspective*, Chicago: University of Chicago Press.

Birch, K. (2017) 'Rethinking value in the bio-economy: Finance, assetization, and the management of value', *Science, Technology, & Human Values*, 42(3): 460–490.

Birch, K. (2020) 'Technoscience rent: Toward a theory of *rentiership* for technoscientific capitalism', *Science, Technology, & Human Values*, 45(1): 3–33. https://doi.org/10.1177/0162243919829567

Birch, K. and Muniesa, F. (2020) 'Introduction: Assetization and technoscientific capitalism', in K. Birch and F. Muniesa (eds) *Assetization: Turning Things into Assets in Technoscientific Capitalism*. Cambridge, MA: MIT Press, pp 1–41. https://doi.org/10.7551/mitpress/12075.001.0001

Blackstone, W. (2016) *Commentaries on the Laws of England*, Oxford: Oxford University Press.

Blok, A., Farías, I. and Roberts, C. (eds) (2019) *The Routledge Companion to Actor-Network Theory*, Abingdon: Routledge.

Blomley, N. (2004) *Unsettling the City: Urban Land and the Politics of Property*, New York: Routledge.

Blomley, N. (2007) 'Making private property: Enclosure, common right and the work of hedges', *Rural History*, 18(1): 1–21.

Blomley, N. (2008) 'Enclosure, common right and the property of the poor', *Social & Legal Studies*, 17(3): 311–331.

Blomley, N. (2013) 'Performing property: Making the world', *Canadian Journal of Law & Jurisprudence*, 26(1): 23–48.

Blomley, N. (2014) 'Disentangling law: The practice of bracketing', *Annual Review of Law and Social Science*, 10(1): 133–148.

Bogner, A., Littig, B. and Menz, W. (eds) (2009) *Interviewing Experts*, London: Palgrave Macmillan UK.

Bollier, D. (2002) 'The enclosure of the academic commons', *Academe*, 88(5): 18–22.

Bonneuil, C. (2006) 'Mendelism, plant breeding and experimental cultures: Agriculture and the development of genetics in France', *Journal of the History of Biology*, 39: 281–308.

Bonneuil, C. and Thomas, F. (2009) *Gènes, pouvoirs et profits: Recherche publique et régimes de production des savoirs de Mendel aux OGM*, Versailles: Quae.

Bonneuil, C. and Thomas, F. (2012) *Semences, une histoire politique: Amélioration des plantes, agriculture et alimentation en France depuis la Seconde Guerre Mondiale*, Paris: Éditions Charles Léopold Mayer.

Bonny, S. (2003) 'Why are most Europeans opposed to GMOs? Factors explaining rejection in France and Europe', *Electronic Journal of Biotechnology*, 6(1): 50–71.

Bonny, S. (2014) 'Taking stock of the genetically modified seed sector worldwide: market, stakeholders, and prices', *Food Security*, 6(4): 525–540.

Bonny, S. (2017) 'Corporate concentration and technological change in the global seed industry', *Sustainability*, 9(1632): 1–25.

Böschen, S., Brandl, B., Gill, B., Schneider, M. and Spranger, P. (2013) 'Innovationsförderung durch geistiges Eigentum? Passungsprobleme zwischen unternehmerischen Wissensinvestitionen und den Schutzmöglichkeiten durch Patente', in E. Grande, D. Jansen, O. Jarren, A. Rip, U. Schimank and p Weingart (eds) *Neue Governance der Wissenschaft*, Bielefeld: Transcript.

Bourdieu, P. (1984) *Distinction: A Social Critique of the Judgement of Taste*, Cambridge, MA: Harvard University Press.

Bourdieu, P. (2008) *The Logic of Practice*, Stanford, CA: Stanford University Press.

Bowker, G. (1992) 'What's in a patent?', in W.E. Bijker and J. Law (eds) *Shaping Technology/building Society: Studies in Sociotechnical Change*, Cambridge, MA: MIT Press, pp 53–67.

brand eins (nd) 'Was wurde aus … der Kartoffelsorte Linda?', *brand eins*, available from https://www.brandeins.de/zugabe/rubriken/was-wurde-aus/der-kartoffelsorte-linda [Accessed: 5 October 2018].

Brandl, B. (2017) *Wissenschaft, Technologieentwicklung und die Spielarten des Kapitalismus: Analyse der Entwicklung von Saatgut in USA und Deutschland*, Wiesbaden: Springer VS.

Brandl, B. and Glenna, L.L. (2017) 'Intellectual property and agricultural science and innovation in Germany and the United States', *Science, Technology, & Human Values*, 42(4): 622–656.

Brandl, B. and Schleissing, S. (eds) (2016) *Biopatente: Saatgut als Ware und als öffentliches Gut*. Baden-Baden: Nomos.

Brandl, B., Paula, K. and Gill, B. (2014) 'Spielarten des Wissenskapitalismus – Die Kommodifizierung von Saatgut in den USA und in Deutschland', *Leviathan*, 42(4): 539–572.

Braun, V. (2020) 'From commodity to asset and back again: Property in the capitalism of varieties', in K. Birch and F. Muniesa (eds) *Assetization: Turning Things into Assets in Technoscientific Capitalism*, Cambridge, MA: MIT Press, pp 203–224.

Braun, V. (2021a) 'Holding on to and letting go of seed: Quasi-commodities and the passage of property', *Journal of Cultural Economy*, 14(3): 306–318.

Braun, V. (2021b) 'Tools of extraction or means of speculation? Making sense of patents in the bioeconomy', in M. Backhouse, R. Lehmann, K. Lorenzen, M. Lühmann, J. Puder, F. Rodríguez et al (eds) *Bioeconomy and Global Inequalities*, Cham: Springer International Publishing, pp 65–84.

Braun, V. (2023) 'The essence of biology: Tomatoes, broccoli and European patents on native plant traits', in J. Bellido and B. Sherman (eds) *Intellectual Property and the Design of Nature*, Oxford: Oxford University Press, pp 49–72.

Braun, V. and Gill, B. (2018) 'Lost in translation: Biofakte auf dem Weg vom Labor ins Patentamt', in B. Gill, F. Torma and K. Zachmann (eds) *Mit Biofakten leben. Sprache und Materialität von Pflanzen und Lebensmitteln*, Baden-Baden: Nomos, pp 129–154.

Braun, V., Brill, S. and Dobeson, A. (2021) 'The mutability of economic things', *Journal of Cultural Economy*, 14(3): 271–279.

Broß, S. (2014) 'Einheitspatent und Einheitliches Patentgericht im europäischen Integrationsprozess – verfassungsrechtliche Perspektive', *Zeitschrift für Geistiges Eigentum / Intellectual Property Journal*, 6(1): 89–105.

Bullard, M. (1990a) 'Ein Exklusivrecht auf profitables Leben', *die tageszeitung*, 23 April 1991, p 11.

Bullard, M. (1990b) 'Müssen Bauern vor der Aussaat für Patente zahlen?', *die tageszeitung*, 8 December 1990, p 11.

Burawoy, M. (2010) *Manufacturing Consent: Changes in the Labor Process Under Monopoly Capitalism*. Chicago: University of Chicago Press.

Butler, J. (2010) 'Performative agency', *Journal of Cultural Economy*, 3(2): 147–161.

Çalışkan, K. and Callon, M. (2009) 'Economization, Part 1: Shifting attention from the economy towards processes of economization', *Economy and Society*, 38(3): 369–398. https://doi.org/10.1080/03085140903020580

Callon, M. (1980) 'Struggles and negotiations to define what is problematic and what is not', in K.D. Knorr, R. Krohn and R. Whitley (eds) *The Social Process of Scientific Investigation*. Dodrecht: D. Reidel Publishing Company, pp 197–219.

Callon, M. (1984) 'Some elements of a sociology of translation: Domestication of the scallops and the fishermen of St Brieuc Bay', *The Sociological Review*, 32(1_suppl): 196–233.

Callon, M. (1994) 'Is science a public good? Fifth Mullins Lecture, Virginia Polytechnic Institute', 23 March 1993, *Science, Technology & Human Values*, 19(4): 395–424.

Callon, M. (1998a) 'Introduction: The embeddedness of economic markets in economics', *The Sociological Review*, 46(S1): 1–57.

Callon, M. (1998b) 'An essay on framing and overflowing: economic externalities revisited by sociology', *The Sociological Review*, 46(S1): 244–269.

Callon, M. (1999) 'Actor-Network Theory – The market test', *The Sociological Review*, 47(S1): 181–195.

Callon, M. (2006) *What Does it Mean to Say that Economics Is Performative?* Paris: Centre de Sociologie de l'Innovation.

Callon, M. (2010) 'Performativity, misfires and politics', *Journal of Cultural Economy*, 3(2): 163–169.

Callon, M. (2021) *Markets in the Making: Rethinking Competition, Goods, and Innovation*, Brooklyn, NY: Zone Books.

Callon, M. and Latour, B. (1997) '"Tu ne calculeras pas!" ou comment symétriser le don et le capital', in A. Caillé (ed) *Le capitalisme aujourd'hui*, Paris: La Découverte, pp 45–70.

Callon, M. and Muniesa, F. (2005) 'Peripheral vision: Economic markets as calculative collective devices', *Organization Studies*, 26(8): 1229–1250.

Callon, M., Millo, Y. and Muniesa, F. (eds) (2007) *Market Devices*. Malden, MA: Blackwell Publishing.

Calvert, J. and Joly, P.-B. (2011) 'How did the gene become a chemical compound? The ontology of the gene and the patenting of DNA', *Social Science Information*, 50(2): 157–177.

Carolan, M.S. (2010) 'The mutability of biotechnology patents: From unwieldy products of nature to independent "object/s"', *Theory, Culture & Society*, 27(1): 110–129.

Carruthers, B.G. and Ariovich, L. (2004) 'The sociology of property rights', *Annual Review of Sociology*, 30(1): 23–46.

Chapman, S. (2018) 'To make one's name famous: Varietal innovation and intellectual property in The Gambia', *American Ethnologist*, 45(4): 482–494.

Chapman, S. (2022) 'The (In)visible labour of varietal innovation', in J. Bangham, X.S. Chacko and S. Kaplan (eds) *Invisible Labour in Modern Science*, London: Rowman & Littlefield International, pp 163–172.

Charles, D. (2001) *Lords of the Harvest. Biotech, Big Money, and the Future of Food*, Cambridge, MA: Perseus Publishing.

Coase, R.H. (1937) 'The nature of the firm', *Economica*, 4(16): 386–405.

Coase, R.H. (1960) 'The problem of social cost', in C. Gopalakrishnan (ed) *Classic Papers in Natural Resource Economics*. London: Palgrave Macmillan UK, pp 87–137.

Coase, R.H. (1974) 'The lighthouse in economics', *Journal of Law & Economics*, 17(2): 357–376.

Cochoy, F. (2004) 'Is the modern consumer a Buridan's donkey? Product packaging and consumer choice', in K.M. Ekström and H. Brembeck (eds) *Elusive Consumption*, Oxford: Berg, pp 205–227.

Cochoy, F. (2007) 'A sociology of market-things: On tending the garden of choices in mass retailing', *The Sociological Review*, 55(2_suppl): 109–129.

Collins, H.M. (1974) 'The TEA set: Tacit knowledge and scientific networks', *Social Studies of Science*, 4(2): 165–185.

Correa, C.M. and Correa, J.I. (2023) 'Manufacturing for export: A TRIPS-consistent pro-competitive exception', in C. Godt and M. Lamping (eds) *A Critical Mind: Hanns Ullrich's Footprint in Internal Market Law, Antitrust and Intellectual Property*, Berlin: Springer, pp 705–731.

Cronon, W. (1991) *Nature's Metropolis: Chicago and the Great West*, New York: W.W. Norton.

Curry, H.A. (2016) *Evolution Made to Order: Plant Breeding and Technological Innovation in Twentieth-Century America*, Chicago: University of Chicago Press.

Dányi, E. (2018) 'The things of the parliament', in J. Brichzin, D. Krichewsky, L. Ringel and J. Schank (eds) *Soziologie der Parlamente: Neue Wege der politischen Institutionenforschung*, Wiesbaden: Springer Fachmedien (Politische Soziologie), pp 267–285.

Darwin, C. (2008) *On the Origin of Species*, Oxford: Oxford University Press.

Daston, L. (2004) 'Type specimens and scientific memory', *Critical Inquiry*, 31(1): 153–182.

Davis, J. (1973) *Land and Family in Pisticci*, London: Athlone Press.

de la Cadena, M. (2010) 'Indigenous cosmopolitics in the Andes: Conceptual reflections beyond "politics"', *Cultural Anthropology*, 25(2): 334–370.

de Laet, M. (2000) 'Patents, travel, space: Ethnographic encounters with objects in transit', *Environment and Planning D: Society and Space*, 18(2): 149–168.

de Rassenfosse, G., Dernis, H. and Boedt, G. (2014) 'An introduction to the Patstat database with example queries', *Australian Economic Review*, 47(3): 395–408.

Demeulenaere, E. (2014) 'A political ontology of seeds: The transformative frictions of a farmers' movement in Europe', *Focaal*, 2014(69): 45–61.

Demsetz, H. (1967) 'Toward a theory of property rights', *The American Economic Review*, 57(2): 347–359.

den Hond, F. (1998) 'On the structuring of variation in innovation processes: A case of new product development in the crop protection industry', *Research Policy*, 27(4): 349–367.

Dickenson, D. (2017) *Property in the Body: Feminist Perspectives*, 2nd ed, Cambridge: Cambridge University Press.

Dickson, D. (1984) *The New Politics of Science*, Oxford: Oxford University Press.

DLZ (2010) 'Viele verdienen mit', *DLZ Praxis*, 10: 36–39.

Dobeson, A. (2019) *Revaluing Coastal Fisheries: How Small Boats Navigate New Markets and Technology*, Cham: Springer.

Dobeson, A. and Kohl, S. (2023) 'The moral economy of land: From land reform to ownership society, 1880–2018', *Socio-Economic Review*, online first: 1–28.

Dobeson, A., Brill, S. and Braun, V. (2023) 'Sorting out economic forms: A field guide to contemporary capitalism', *Distinktion: Journal of Social Theory*. https://doi.org/10.1080/1600910X.2023.2245573.

Doctorow, C. (2008) *Content: Selected Essays on Technology, Creativity, Copyright, and the Future of the Future*, San Francisco: Tachyon Publications.

Doing, P. (2008) 'Give me a laboratory and I will raise a discipline: The past, present and future politics of laboratory studies in STS', in E.J. Hackett, O. Amsterdamska, M.E. Lynch and J. Wajcman (eds) *The Handbook of Science and Technology Studies*, 3rd ed, Cambridge, MA: MIT Press.

Dreier, T. (2023) 'Caught between post- and neo-colonialism: IP for traditional knowledge, traditional cultural expressions and indigenous resources', in C. Godt and M. Lamping (eds) *A Critical Mind: Hanns Ullrich's Footprint in Internal Market Law, Antitrust and Intellectual Property*, Berlin: Springer, pp 785–806.

du Gay, P. and Pryke, M. (eds) (2002) *Cultural Economy: Cultural Analysis and Commercial Life. Workshop on Cultural Economy*, London: SAGE.

Dunlop, H. (2017) 'European Patent Office attacks its examination backlog', *Maucher Jenkins*, [online] 27 November 2017, available from: https://www.maucherjenkins.com/commentary/european-patent-office-attacks-its-examination-backlog [Accessed: 21 April 2023].

Dupre, J. (2004) 'Understanding contemporary genomics', *Perspectives on Science*, 12(3): 320–338.

Dupret, B., Lynch, M. and Berard, T. (eds) (2015) *Law at Work: Studies in Legal Ethnomethods*, Oxford: Oxford University Press.

Dutfield, G. (2003) 'Bioprospecting: Legitimate research or "biopiracy"?', *Policy Briefs, Sci. & Dev. Network*, [online] 26 May 2003, available from: http://www.scidev.net/en/agriculture-and-environment/bioprospecting/policy-briefs/bioprospecting-legitimateresearch-or-biopiracy--1.html [Accessed: 18 September 2018].

Dutfield, G. (2004) *Intellectual Property, Biogenetic Resources, and Traditional Knowledge*, London: Earthscan.

Dutfield, G. (2017) 'TK unlimited: The emerging but incoherent international law of traditional knowledge protection', *The Journal of World Intellectual Property*, 20(5–6): 144–159.

Dutfield, G. (2018) 'Farmers, innovation and intellectual property: Current trends and their consequences for food security', in F. Girard and C. Frison (eds) *The Commons, Plant Breeding and Agricultural Research: Challenges for Food Security and Agrobiodiversity*, New York: Routledge, pp 21–38.

Dutfield, G. and Suthersanen, U. (2005) 'Harmonisation or differentiation in intellectual property protection? The lessons of history', *Prometheus*, 23(2): 131–147.

Earle, T. (2017) 'Property in prehistory', in M. Graziadei and L.D. Smith (eds) *Comparative Property Law*, Cheltenham: Edward Elgar Publishing, pp 3–25.

Eisenberg, R.S. (2006) 'Biotech patents: Looking backward while moving forward', *Nature Biotechnology*, 24(3): 317–319.

Elkin-Koren, N. (2007) 'Making room for consumers under the DMCA', *Berkeley Technology Law Journal*, 22(3): 1119–1155.

Elmore, B.J. (2021) *Seed Money: Monsanto's Past and our Food Future*, New York: W.W. Norton & Company.

EPO (2016) EPO stays proceedings in certain biotechnology cases, [online] 12 December 2016, available from: https://www.epo.org/news-issues/news/2016/20161212.html [Accessed: 16 October 2018].

EPO (2017) EPO clarifies practice in the area of plant and animal patents, [online] 29 June 2017, available from: https://www.epo.org/news-issues/news/2017/20170629.html [Accessed: 28 August 2017].

Ernst, H. (2017) 'Intellectual property as a management discipline', *Technology & Innovation*, 19(2): 481–492.

Errygers, G. and Cunliffe, J. (eds) (2012) *Inherited Wealth, Justice and Equality*, London: Routledge.

European Commission (2016) 'Commission Notice on certain articles of Directive 98/44/EC of the European Parliament and of the Council on the legal protection of biotechnological inventions', *Official Journal of the European Union*, C411: 3–14.

Euroseeds (2019) 'Intellectual Property protection for plant-related inventions in Europe', Euroseeds, available from: https://euroseeds.eu/app/uploads/2019/07/12.0100-Euroseeds-position-Intellectual-Property.pdf [Accessed: 10 December 2023].

Feenan, D. (ed) (2013) *Exploring the 'Socio' of Socio-Legal Studies*, Basingstoke: Palgrave Macmillan.

Fitzgerald, D. (1990) *The Business of Breeding: Hybrid Corn in Illinois, 1890–1940*, Ithaca: Cornell University Press.

Fitzgerald, D. (2003) *Every Farm a Factory: The Industrial Ideal in American Agriculture*, New Haven, CT: Yale University Press.

FitzGerald, W.F. (1895) 'The lost property office', *The Strand*, 10: 641–653.

Fleck, L. (1981) *Genesis and Development of a Scientific Fact*, Chicago: University of Chicago Press.

Forelle, M.C. (2017) Copyright and Car Modding: The Conflict Over Vehicle Ownership and Vehicle Software Copyright. Presented at the 4S Meeting, Boston, MA, 2 September 2017.

Foucault, M. (2002) *Archaeology of Knowledge*. London: Routledge.

Foucault, M. (2005) *Society Must Be Defended: Lectures at the Collège de France, 1975–76*, 1st ed, New York: Picador.

Fourcade, M., Ollion, E. and Algan, Y. (2015) 'The superiority of economists', *Journal of Economic Perspectives*, 29(1): 89–114.

Fowler, C. (1994) *Unnatural Selection: Technology, Politics, and Plant Evolution*, Langhorne, PA: Gordon and Breach.

Fowler, C. (2000) 'The Plant Patent Act of 1930: A sociological history of its creation', *Journal of the Patent & Trademark Office Society*, 82: 621–644.

Fox Keller, E. (1983) *A Feeling for the Organism: The Life and Work of Barbara McClintock*, San Francisco: W.H. Freeman.

Fox Keller, E. (2002) *The Century of the Gene*, Cambridge, MA: Harvard University Press.

Fox Keller, E. (2005) 'The century beyond the gene', *Journal of Biosciences*, 30(1): 3–10.

Frison, C. and van Zimmeren, E. (2021) 'Patent thicket', in M. Cornu, F. Orsi and J. Rochfeld (eds) *Dictionnaire des biens communs*, Paris: puf, pp 1301–1305.

Fromm, E. (2005) *To Have or To Be?* London: Continuum.

Fuller, L.L. (1970) *Legal Fictions*, Palo Alto: Stanford University Press.

Fullilove, C. (2017) *The Profit of the Earth: The Global Seeds of American Agriculture*, Chicago: University of Chicago Press.

Galushko, V. and Gray, R. (2013) *Privatization of Crop Breeding in the UK: Lessons for Other Countries*. 87th Annual Conference of the Agricultural Economics Society, Warwick.

Gibson-Graham, J.K. (2006) *The End of Capitalism (As We Knew It): A Feminist Critique of Political Economy*, Minneapolis: University of Minnesota Press.

Gill, B. (2003) *Streitfall Natur: Weltbilder in Technik- und Umweltkonflikten.* Wiesbaden: Westdeutscher Verlag.

Gill, B. and Brandl, B. (2014) 'Legitimität von Sortenschutz und Sortenzulassung aus soziologischer Sicht', in A. Metzger (ed) *Rechtsschutz von Pflanzenzüchtungen. Eine kritische Bestandsaufnahme des Sorten-, Patent- und Saatgutrechts*, Tübingen: Mohr Siebeck, pp 163–186.

Gill, B., Brandl, B., Böschen, S. and Schneider, M. (2012) 'Autorisierung. Eine wissenschafts- und wirtschaftssoziologische Perspektive auf geistiges Eigentum', *Berliner Journal für Soziologie*, 22(3): 407–440.

Gill, B., Torma, F. and Zachmann, K. (eds) (2018) *Mit Biofakten Leben. Sprache und Materialität von Pflanzen und Lebensmiteln*. Baden-Baden: Nomos.

Girard, F. (2015) "'Though the treasure of nature's germens tumble all together": The EPO and patents on native traits or the bewitching powers of ideologies', *Prometheus*, 33(1): 43–65.

Girard, F. and Frison, C. (eds) (2018) *The Commons, Plant Breeding and Agricultural Research: Challenges for Food Security and Agrobiodiversity*, New York: Routledge.

Godin, B. (2017) *Models of Innovation: The History of an Idea*, Cambridge, MA: MIT Press.

Gold, E.R. (1996) *Body Parts: Property Rights and the Ownership of Human Biological Materials*, Washington, DC: Georgetown University Press.

Goodman, M.M. (2002) 'New sources of germplasm: lines, transgenes, and breeders', in J.M. Martinez, F. Rincon, and G. Martinez (eds) *Memoria Congresso Nacional de Fitogenetica*, Saltillo, Coah.: Universidad Autónoma Agraria Antonio Narro, pp 28–41.

Gouldner, A.W. (1970) *The Coming Crisis of Western Sociology*, London: Heinemann.

Graeber, D. (2011) 'Consumption', *Current Anthropology*, 52(4): 489–511.

Graff, G.D., Cullen, S.E., Bradford, K.J., Zilberman, D. and Bennett, A.B. (2003) The public–private structure of intellectual property ownership in agricultural biotechnology. *Nature Biotechnology*, 21(9): 989–995.

Graham, S.J.H., Merges, R.P., Samuelson, p and Sichelman, T. (2009) 'High technology entrepreneurs and the patent system: Results of the 2008 Berkeley Patent Survey', *Berkeley Technology Law Journal*, 24(4): 1255–1327.

Graziadei, M. (2017) 'The structure of property ownership and the common law/civil law divide', in M. Graziadei and L. Smith (eds) *Comparative Property Law.* Cheltenham: Edward Elgar Publishing, pp 71–99.

Graziadei, M. and Smith, L.D. (eds) (2017) *Comparative Property Law. Global Perspectives*, Cheltenham: Edward Elgar Publishing.

Gregory, C.A. (1982) *Gifts and Commodities*, London: Academic Press.

Greimas, A.J. (1983) *Structural Semantics: An Attempt at a Method*, Lincoln, NB: University of Nebraska Press.

Grey, T.C. (1980) 'The disintegration of property', *Nomos*, 22: 69–85.

Gusterson, H. (1997) 'Studying up revisited', *PoLAR: Political and Legal Anthropology Review*, 20(1): 114–119.

Hacking, I. (1983) *Representing and Intervening: Introductory Topics in the Philosophy of Natural Science*, Cambridge: Cambridge University Press.

Hacking, I. (1999) *The Social Construction of What?*, Cambridge, MA: Harvard University Press.

Hall, B.H. and Ziedonis, R.H. (2001) 'The patent paradox revisited: An Empirical study of patenting in the U.S. semiconductor industry, 1979–1995', *The RAND Journal of Economics*, 32(1): 101–128.

Hallonsten, O. (2016) *Big Science Transformed: Science, Politics and Organization in Europe and the United States.* Basingstoke: Palgrave Macmillan.

Handelsblatt (1987) 'Biotechnologie: Die Industrie braucht verlässliche, juristische Grundlagen', *Handelsblatt,* 23 March 1987, p 19.

Hann, C. (ed) (1998a) *Property Relations. Renewing the Anthropological Tradition,* Cambridge: Cambridge University Press.

Hann, C. (1998b) 'Introduction: The embeddedness of property', in C. Hann (ed) *Property Relations: Renewing the Anthropological Tradition,* Cambridge: Cambridge University Press, pp 1–47.

Hann, C. (2005) 'Property', in J.G. Carrier (ed) *A Handbook of Economic Anthropology,* Cheltenham: Edward Elgar Publishing, pp 110–124.

Hann, C. (2006) 'Propertization und ihre Gegentendenzen: Beispiele aus ländlichen Gebieten Europas', *Comparativ,* 16(5/6): 84–98.

Hann, C. (2007) 'A new double movement? Anthropological perspectives on property in the age of neoliberalism', *Socio-Economic Review,* 5(2): 287–318.

Hartung, F. and Schiemann, J. (2014) 'Precise plant breeding using new genome editing techniques: opportunities, safety and regulation in the EU', *The Plant Journal,* 78(5): 742–752.

Harwood, J. (2012) *Europe's Green Revolution and Others since: The Rise and Fall of Peasant-Friendly Plant Breeding,* London: Routledge.

Hayden, C. (2003) *When Nature Goes Public: The Making and Unmaking of Bioprospecting in Mexico,* Princeton, NJ: Princeton University Press.

Hayden, C. (2011) 'Bioprospecting's representational dilemma', in S.G. Harding (ed) *The Postcolonial Science and Technology Studies Reader,* Durham: Duke University Press, pp 343–363.

Head, L., Atchison, J. and Gates, A. (2012) *Ingrained: A Human Bio-geography of Wheat,* Abingdon: Routledge.

Heeren, A. (2016) 'Commercialization of biodiversity: The regulation of bioprospecting in Ecuador', *Fiar,* 9(2): 94–117.

Heeren, A. (2017) 'Globale Bioökonomien. Gesellschaftliche Naturverhältnisse im Kontext der Bioprospektion in Namibia', in S. Lessenich (ed) *Geschlossene Gesellschaften. Verhandlungen des 38. Kongresses der Deutschen Gesellschaft für Soziologie in Bamberg 2016,* Essen: Deutsche Gesellschaft für Soziologie.

Heinacher, P. (1988) 'Bedeutung der Gentechnologie macht Anpassung des Patentrechts notwendig', *Handelsblatt,* 24 August 1988, p 4.

Heisey, P.W., King, J.L. and Rubenstein, K.D. (2005) 'Patterns of public-sector and private-sector patenting in agricultural biotechnology', *AgBioForum,* 8(2/3): 73–82.

Heitz, A. (1991) 'The history of the UPOV Convention and the rationale for plant breeders' rights', in UPOV (ed) *Seminar on the Nature of and Rationale for the Protection of Plant Varieties under the UPOV Convention,* Geneva: UPOV, pp 19–42.

Helfrich, S. and Bollier, D. (2021) 'Commoning', in M. Cornu, F. Orsi and J. Rochfeld (eds) *Dictionnaire des biens communs*, 2nd ed, Paris: puf, pp 222–228.

Heller, M.A. and Eisenberg, R.S. (1998) 'Can patents deter innovation? The anticommons in biomedical research', *Science*, 280(5364): 698–701.

Hendriks, A. (2017) *Tomaten: Die wahre Identität unseres Frischgemüses. Eine Reportage*, Berlin: be.bra verlag.

Hicks, D.J. (2017) 'Genetically modified crops, inclusion, and democracy', *Perspectives on Science*, 25(4): 488–520.

Hiscock, R., Kearns, A., MacIntyre, S. and Ellaway, A. (2001) 'Ontological security and psycho-social benefits from the home: Qualitative evidence on issues of tenure', *Housing, Theory and Society*, 18(1–2): 50–66.

Hoeyer, K. (2007) 'Person, patent and property: A critique of the commodification hypothesis', *BioSocieties*, 2(3): 327–348.

Hoeyer, K. (2009) 'Tradable body parts? How bone and recycled prosthetic devices acquire a price without forming a "market"', *BioSocieties*, 4(2–3): 239–256.

Hohfeld, W.N. (1920) *Fundamental Legal Conceptions as Applied in Judicial Reasoning and Other Legal Essays*, New Haven: Yale University Press.

Hume, D. (1960) *A Treatise on Human Nature*, Oxford: Clarendon Press.

Hummel, P., Braun, M. and Dabrock, P. (2021) 'Own data? Ethical reflections on data ownership', *Philosophy & Technology*, 34(3): 545–572.

Ipsen, A. (2016) 'Manufacturing a natural advantage: Capturing place-based technology rents in the genetically modified corn seed industry', *Environmental Sociology*, 2(1): 41–52.

Irle, M. (2005) '"Spreche ich eigentlich kein Deutsch mehr?"', *brand eins*, [online] 2005, available from https://www.brandeins.de/magazine/brand-eins-wirtschaftsmagazin/2005/kommunikation/spreche-ich-eigentlich-kein-deutsch-mehr [Accessed: 8 October 2018].

ISAAA (2018) *Global Status of Commercialized Biotech/GM Crops in 2017: Biotech Crop Adoption Surges as Economic Benefits Accumulate in 22 Years*, Ithaca: ISAAA.

Jaffe, A.B. and Lerner, J. (2011) *Innovation and Its Discontents: How Our Broken Patent System Is Endangering Innovation and Progress, and What to Do about It*, Princeton, NJ: Princeton University Press.

James, W. (1898) 'Philosophical conceptions and practical results', *University Chronicle*, 1(4): 287–310.

Janis, M.D. (2017) 'Nonobvious plants', in D. Matthews and H. Zech (eds) *Research Handbook on Intellectual Property and the Life Sciences*, Cheltenham: Edward Elgar, pp 160–178.

Jasanoff, S. (1997) *Science at the Bar: Law, Science, and Technology in America*, 2nd ed, Cambridge, MA: Harvard University Press.

Jasanoff, S. (2003) 'Breaking the waves in science studies: Comment on H.M. Collins and Robert Evans, "The Third Wave of Science Studies"', *Social Studies of Science*, 33(3): 389–400.

Jasanoff, S. (2007) *Designs on Nature: Science and Democracy in Europe and the United States*, Princeton, NJ: Princeton University Press.

Johnson, J. (1988) 'Mixing humans and nonhumans together: The sociology of a door-closer', *Social Problems*, 35(3): 298–310. https://doi.org/10.2307/800624

Johnson, S.D., Sidebottom, A. and Thorpe, A. (2008) *Bicycle Theft*, Albany: Center for Problem-Oriented Policing.

Kang, H.Y. (2015) 'Patent as credit. When intellectual property becomes speculative', *Radical Philosophy*, 194: 29–37.

Kang, H.Y. (2019) 'Ghosts of inventions: Patent law's digital mediations', *History of Science*, 57(1): 38–61.

Kang, H.Y. (2020) 'Patents as assets: Intellectual property rights as market subjects and objects', in K. Birch and F. Muniesa (ed) *Turning Things into Assets*. Cambridge, MA: MIT Press, pp 45–74.

Kang, H.Y. (2023) 'Patents as capitalist aesthetic forms', *Law and Critique*. https://doi.org/10.1007/s10978-023-09349-2

Kang, H.Y. and Kendall, S. (2020) 'Legal materiality', in S. Stern, M. Del Mar and B. Meyler (eds) *The Oxford Handbook of Law and Humanities*, 1st ed, Oxford: Oxford University Press, pp 20–37.

Karpik, L. (2010) *Valuing the Unique: The Economics of Singularities*, Princeton, NJ: Princeton University Press.

Kautsky, K. (1988) *The Agrarian Question*, London: Zwan Publications.

Kempf, H. (2016) *Weizenzüchtung bei der Secobra Saatzucht im Spannungsfeld zwischen Wettbewerb, Landwirtschaft, Handel und Verbraucher*, presentation, Department für Nutzpflanzenwissenschaften, Universität Göttingen.

Kevles, D.J. (1994) 'Ananda Chakrabarty wins a patent: Biotechnology, law, and society, 1972–1980', *Historical Studies in the Physical and Biological Sciences*, 25(1): 111–135.

Kevles, D.J. (2002) 'Of mice and money: The story of the world's first animal patent', *Daedalus*, 131(2): 78–88.

Kevles, D.J. (2007) 'Patents, protections, and privileges: The establishment of intellectual property in animals and plants', *Isis*, 98(2): 323–331.

Kevles, D.J. (2011) 'New blood, new fruits: protections for breeders and originators, 1789–1930', in M. Biagioli, P. Jaszi and M. Woodmansee (eds) *Making and Unmaking Intellectual Property*, Chicago: University of Chicago Press, pp 253–268.

Klevnäs, A. and Hedenstierna-Jonson, C. (eds) (2015) *Own and be Owned: Archaeological Approaches to the Concept of Possession*, Stockholm: Department of Archaeology and Classical Studies, Stockholm University.

Kloppenburg, J. (2004) *First the Seed. The Political Economy of Plant Biotechnology*, Madison: University of Wisconsin Press.

Kloppenburg, J. (2014) 'Re-purposing the master's tools: The open source seed initiative and the struggle for seed sovereignty', *The Journal of Peasant Studies*, 41(6): 1225–1246.

Knorr, K.D. and Knorr, D.W. (1978) *From Scenes to Scripts: On the Relationship between Laboratory Research and Published Paper in Research*, Vienna: Institute for Advanced Studies.

Kock, M.A. and Gould, C. (2014) 'Patents on plants: A tool or threat for sustainable agriculture? The role of intellectual property rights on plant innovations', in *How the Private and the Public Sectors Use Intellectual Property to Enhance Agricultural Productivity*, Geneva: WIPO, pp 94–104.

Kock, M.A. and ten Have, F. (2016) 'The "International Licensing Platform – Vegetables": A prototype of a patent clearing house in the life science industry', *Journal of Intellectual Property Law & Practice*, 11(7): 496–515.

Kotschi, J. and Horneburg, B. (2018) 'The Open Source Seed Licence: A novel approach to safeguarding access to plant germplasm', *PLOS Biology*, 16(10): e3000023.

Kramer, M. (1980) *Three Farms: Making Milk, Meat and Money from the American Soil*, Boston: Little & Brown.

Krasikov, H.N. (2022) 'The NFT boom and bust', *Journal of Popular Music Studies*, 34(4): 39–60.

Krieger, A. (1989) 'Aktuelle Entwicklungen auf dem Gebiet des internationalen Schutzes des geistigen Eigentums', *Der Betrieb*, 1989(17): 865–870.

Kupferschmidt, K. (2018) 'Biologists raise alarm over changes to biopiracy rules', *Science*, 361(6397): 14.

Lametti, D. (2013) 'The concept of the anticommons: Useful, or ubiquitous and unnecessary?', in H. Howe and J. Griffiths (eds) *Concepts of Property in Intellectual Property Law*, Cambridge: Cambridge University Press, pp 232–257.

Landecker, H. (2016) 'Antibiotic resistance and the biology of history', *Body & Society*, 22(4): 19–52.

Landes, W.M. and Posner, R.A. (2003) *The Economic Structure of Intellectual Property Law*, Cambridge, MA: Harvard University Press.

Langbein, C. (1995) 'Züchtungsfortschritt muss entlohnt werden', *Ernährungsdienst*, 86: P002.

Lassoued, R., Phillips, P.W.B., Smyth, S.J. and Hesseln, H. (2019) 'Estimating the cost of regulating genome edited crops: Expert judgment and overconfidence', *GM Crops & Food*, 10(1): 44–62.

Latour, B. (1984) 'Le dernier des capitalistes sauvages: Interview d'un biochimiste', *Fundamenta Scientiae*, 4(3/4): 301–327.

Latour, B. (1987) *Science in Action: How to Follow Scientists and Engineers Through Society*, Cambridge, MA: Harvard University Press.

Latour, B. (1990) 'Technology is society made durable', *The Sociological Review*, 38(S1): 103–131.

Latour, B. (1993) *The Pasteurization of France*, Cambridge, MA: Harvard University Press.

Latour, B. (1996) 'Do scientific objects have a history? Pasteur and Whitehead in a Bath of Lactic Acid', *Common Knowledge*, 5(1): 76–91.

Latour, B. (2005) *Reassembling the Social: An Introduction to Actor-Network-Theory*, Oxford: Oxford University Press.

Latour, B. (2010) *The Making of Law: An Ethnography of the Conseil d'Etat*, Cambridge: Polity Press.

Latour, B. (2013) *An Inquiry into Modes of Existence: An Anthropology of the Moderns*, Cambridge, MA: Harvard University Press.

Latour, B. and Woolgar, S. (1986) *Laboratory Life. The Construction of Scientific Facts*, Princeton, NJ: Princeton University Press.

Lawson, C. and Rourke, M. (2016) 'Open access DNA, RNA and amino acid sequences: The consequences and solutions for the international regulation of access and benefit sharing', *Journal of Law and Medicine*, 24: 96–118.

Lebrecht, T. and Meienberg, F. (2014) 'Private claims on nature: No to Syngenta's patent on peppers', *No Patents on Seeds*, [online] February 2014, available from: https://www.no-patents-on-seeds.org/sites/default/files/2019-01/2014_Brochure_No_to_Syngenta_s_patent_on_peppers.pdf [Accessed: 13 December 2023].

Lemley, M.A. (2000) 'Reconceiving patents in the age of venture capital', *Journal of Small and Emerging Business Law*, 4: 137–148.

Leskien, D. (1991) 'Genfer Schlussverkauf im Saatgutmarkt', *die tageszeitung*, 5 March 1991, p 9.

Lewis, S. (2019) *Full Surrogacy Now: Feminism against Family*. London: Verso.

Llewelyn, M. and Adcock, M. (2006) *European Plant Intellectual Property*, Oxford: Hart Publishing.

Locke, J. (2003) *Two Treatises of Government: And a Letter Concerning Toleration*, New Haven, CT: Yale University Press.

Loick, D. (2023) *The Abuse of Property*, Cambridge, MA: MIT Press.

Long, C. (2002) 'Patent signals', *The University of Chicago Law Review*, 69: 625–679.

Lurquin, P.F. (2001) *The Green Phoenix: A History of Genetically Modified Plants*, New York: Columbia University Press.

Lusser, M., Parisi, C., Plan, D. and Rodríguez-Cerezo, E. (2012) 'Deployment of new biotechnologies in plant breeding', *Nature Biotechnology*, 30(3): 231–239.

Machlup, F. (1958) *An Economic Review of the Patent System: Study of the Subcommittee on Patents, Trademarks, and Copyrights of the Committee on the Judiciary, U.S. Senate, 85th Congress, Second Session*, Washington, DC: US Government Printing Office.

MacKenzie, D.A. (2006) *An Engine, Not a Camera: How Financial Models Shape Markets*, Cambridge, MA: MIT Press.

MacKenzie, D.A. (2009) *Material Markets. How Economic Agents are Constructed*. Oxford: Oxford University Press.

MacKenzie, D.A., Muniesa, F. and Siu, L. (eds) (2007) *Do Economists Make Markets? On the Performativity of Economics*, Princeton, NJ: Princeton University Press.

Macpherson, C.B. (1978) *Property. Mainstream and Critical Positions*, Oxford: Blackwell.

Macpherson, C.B. (1990) *The Political Theory of Possessive Individualism: Hobbes to Locke*, Oxford: Oxford University Press.

Madero, M. (2010) *Tabula Picta. Painting and Writing in Medieval Law*. Philadelphia: University of Pennsylvania Press.

Maharaj, N. and Dorren, G. (1995) *The Game of the Rose. The Third World in the Global Flower Trade*. Utrecht, Netherlands: International Books.

Maine, H.S. (1908) *Ancient Law: Its Connection with the Early History of Society and its Relation to Modern Ideas*, London: John Murray.

Malinowski, B. (1984) *Argonauts of the Western Pacific: An Account of Native Enterprise and Adventure in the Archipelagoes of Melanesian New Guinea*. Prospect Heights, IL: Waveland Press.

Mammana, I. (2014) *Concentration of Market Power in the EU Seed Market*, The Greens/EFA Group in the European Parliament, available from: http://greens-efa-service.eu/concentration_of_market_power_in_EU_see_market/files/assets/common/downloads/publication.pdf [Accessed: 10 December 2023].

Marx, K. (1991) *The Capital*. London: Penguin Books.

Marzluff, J.M. and Angell, T. (2005) *In the Company of Crows and Ravens*. New Haven: Yale University Press.

Maurer, B. and Schwab, G. (eds) (2006) *Accelerating Possession: Global Futures of Property and Personhood*, New York: Columbia University Press.

Mauss, M. (2002) *The Gift: The Form and Reason for Exchange in Archaic Societies*. London: Routledge.

May, C. (2000) *A Global Political Economy of Intellectual Property Rights: The New Enclosures?*, London: Routledge.

McDougall, P. (2011) *The Cost and Time Involved in the Discovery, Development and Authorisation of a New Plant Biotechnology Derived Trait*, Midlothian: Phillips McDougall.

Mercier, S. (2021) 'Enclosure de la connaissance (approche pratique)', in M. Cornu, F. Orsi and J. Rochfeld (eds) *Dictionnaire des biens communs*, 2nd ed, Paris: puf, pp 562–565.

Merton, R.K. (1974) *The Sociology of Science: Theoretical and Empirical Investigations*, Chicago: University of Chicago Press.

Metzger, A. (2016) 'Patents on tomatoes and broccoli: Legal positivists at work', *IIC – International Review of Intellectual Property and Competition Law*, 47(5): 515–516.

Metzger, A. and Zech, H. (eds) (2006), *Sortenschutzrecht: SortG, GSortV, PatG, EPÜ: Kommentar*, München: C.H. Beck.

Miller, D. (2002) 'Turning Callon the right way up', *Economy and Society*, 31(2): 218–233.

Mirowski, P. (2011) *Science-Mart: Privatizing American Science*, Cambridge, MA: Harvard University Press.

Mirowski, P. (2012) 'The modern commercialization of science is a passel of Ponzi schemes', *Social Epistemology*, 26(3–4): 285–310.

Misa, T.J. (1992) 'Controversy and closure in technological change: Constructing "steel"', in W.E. Bijker and J. Law (eds) *Shaping Technology/ Building Society: Studies in Sociotechnical Change*, Cambridge, MA: MIT Press, pp 109–139.

Monhait, J. (2013) 'Baseball arbitration: An ADR success', *Harvard Journal of Sports and Entertainment Law*, 4(1): 105–144.

Moore, K.A. (2005) 'Worthless patents', *Berkeley Technology Law Journal*, 30(4): 1527–1552.

Morozov, E. (2022) 'Critique of techno-feudal reason', *New Left Review*, 133/134: 89–126.

Moskowitz, M. (2006) 'Broadcasting seeds on the American landscape', in E.H. Brown, C. Gudis and M. Moskowitz (eds) *Cultures of Commerce: Representation and American Business Culture, 1877–1960*, Basingstoke: Palgrave Macmillan, pp 9–26.

Moskowitz, M. (2008) 'The limits of globalization? The horticultural trades in postbellum America', in A. Nützenadel and F. Trentmann (eds) *Food and Globalization: Consumption, Markets and Politics in the Modern World*, Oxford: Berg, pp 57–74.

Müller, B. (2006) 'Infringing and trespassing plants: Patented seeds at dispute in Canada's courts', *Focaal*, 48: 83–98.

Müller, B. (2007) *Disenchantment with Market Economics: East Germans and Western Capitalism*, New York: Berghahn Books.

Müller, B. (2013) 'The Canadian Wheat Board struggle: Taking freedom and democracy to market', in C. Counihan and V. Siniscalchi (eds) *Food Activism: Agency, Democracy and Economy*, London: Berg Publishers, pp 129–142.

Müller, B. (2015) 'Fools gold on the prairies: Ontologies, farmers and their seeds', *Tsantsa*, 20: 61–73.

Müller-Wille, S. and Rheinberger, H.-J. (2012) *A Cultural History of Heredity*, Chicago: University of Chicago Press.

Muniesa, F. (2008) 'Attachment and detachment in the economy', in P. Redman (ed) *Attachment: Sociology and Social Worlds*, Manchester: Manchester University Press, pp 111–142.

Muniesa, F. (2011) 'A flank movement in the understanding of valuation', *The Sociological Review*, 59(s2): 24–38.

Muniesa, F. (2015) *The Provoked Economy: Economic Reality and the Performative Turn*, London: Routledge.

Muniesa, F., Millo, Y. and Callon, M. (2007) 'An introduction to market devices', *The Sociological Review*, 55(2): 1–12.

Nader, L. (1972) 'Up the anthropologist: Perspectives gained from studying up', in D. Hymes (ed) *Reinventing Anthropology*, New York: Pantheon Books, pp 284–311.

Navaro-Yashin, Y. (2009) 'Affective spaces, melancholic objects: Ruination and the production of anthropological knowledge', *The Journal of the Royal Anthropological Institute*, 15(1): 1–18.

Netz, R. (2009) *Barbed Wire: An Ecology of Modernity*, Middletown, CT: Wesleyan University Press.

Nichols, R. (2020) *Theft is Property! Dispossession & Critical Theory*, Durham: Duke University Press.

Nickl, U., Hartl, L. and Herz, M. (2014) 'Erträge steigern mit Hybridsorten?', in LfL Bayern (ed) *Ackerbau - mit hohen Erträgen erfolgreich wirtschaften*, Freising-Weihenstephan: LfL Bayern, pp 33–40.

Nicol, D., Nielsen, J. and Dawkins, V. (2018) D'Arcy v Myriad genetics: The impact of the high court's decision on the cost of genetic testing in Australia. Centre for Law and Genetics, Occasional Papers nr 9.

No Patents on Seeds! (nd) What is the problem?, *No Patents on Seeds!* Available from: https://www.no-patents-on-seeds.org/en/background/problem [Accessed: 9 December 2023].

North, D.C. (1981) *Structure and Change in Economic History*, New York: W.W. Norton.

Orsi, F. (2021) 'Faisceau de droits (bundle of rights)', in M. Cornu, F. Orsi and J. Rochfeld (eds) *Dictionnaire des biens communs*, 2nd ed, Paris: puf, pp 597–602.

Ostrom, E. (1990) *Governing the Commons: The Evolution of Institutions for Collective Action*, Cambridge: Cambridge University Press.

Ostrom, E. (2005) *Understanding Institutional Diversity*, Princeton, NJ: Princeton University Press.

Ostrom, E., Gardner, R.J. and Walker, J. (1994) *Rules, Games, and Common-Pool Resources*, Ann Arbor: University of Michigan Press.

Ouma, S. (2020) *Farming as Financial Asset: Global Finance and the Making of Institutional Landscapes*, Newcastle upon Tyne: Agenda Publishing.

Page, J. (2017) *Property Diversity and its Implications*, London: Routledge.

Pallauf, M. (2018) *Planning and Decision-making by Seed Trading Companies for Winter Wheat (Triticum aestivum) Multiplication: A Mixed-Methods Analysis for Bavaria*, MA thesis, Weihenstephan: TU Munich.

Pannhorst, K. (2016) 'Verpacken, verkaufen, verschenken: Hans Sauters entomologische Praktiken zwischen Formosa und Europa, 1902–1914', *Berichte zur Wissenschaftsgeschichte*, 39(3): 230–244.

Parance, B. (2021) 'Ressources génétiques', in M. Cornu, F. Orsi, and J. Rochfeld (eds) *Dictionnaire des biens communs*, Paris: puf, pp 1149–1153.

Parry, B. (2008) 'Entangled exchange: Reconceptualising the characterisation and practice of bodily commodification', *Geoforum*, 39(3): 1133–1144.

Parthasarathy, S. (2017) *Patent Politics: Life Forms, Markets, and the Public Interest in the United States and Europe*, Chicago: University of Chicago Press.

Peavey, T.M. (2014) 'Bowman v. Monsanto: Bowman, the producer and the end user', *Berkeley Technology Law Journal*, 29: 465–492.

Penner, J.E. (1996) 'The "bundle of rights" picture of property', *UCLA Law Review*, 43: 711–820.

Perzanowski, A. and Schultz, J. (2018) *The End of Ownership: Personal Property in the Digital Economy*, Cambridge, MA: MIT Press.

Peschard, K. and Randeria, S. (2020) '"Keeping seeds in our hands": The rise of seed activism', *The Journal of Peasant Studies*, 47(4): 613–647.

Peukert, A. (2021) *A Critique of the Ontology of Intellectual Property Law*, Cambridge: Cambridge University Press.

Pickering, A. (1995) *The Mangle of Practice: Time, Agency, and Science*. Chicago: University of Chicago Press.

Pigou, A.C. (1932) *The Economics of Welfare*, London: Macmillan and Co.

Piketty, T. (2014) *Capital in the Twenty-First Century*, Cambridge, MA: Belknap Press of Harvard University Press.

Pistor, K. (2019) *The Code of Capital: How the Law Creates Wealth and Inequality*, Princeton, NJ: Princeton University Press.

Plesner, U. (2011) 'Studying sideways: Displacing the problem of power in research interviews with sociologists and journalists', *Qualitative Inquiry*, 17(6): 471–482.

Pollack Petchesky, R. (1995) 'The body as property: A feminist re-vision', in F. Ginsburg and R. Rapp (eds) *Conceiving the New World Order: The Global Politics of Reproduction*, Berkeley: University of California Press, pp 387–406.

Posner, R.A. (1986) *Economic Analysis of Law*, 3rd ed, Boston, MA: Little, Brown and Company.

Pottage, A. (1998) 'Instituting property', *Oxford Journal of Legal Studies*, 18(2): 331–344.

Pottage, A. (2004) 'Who owns academic knowledge?', *Cambridge Anthropology*, 24(1): 1–20.

Pottage, A. (2006) 'Too much ownership: Bio-prospecting in the age of synthetic biology', *BioSocieties*, 1(2): 137–158.

Pottage, A. (2011) 'Biotechnology as environmental regulation', in A. Philippopoulos-Mihalopoulos (ed) *Law and Ecology*, Abingdon: Routledge, pp 105–125.

Pottage, A. and Sherman, B. (2007) 'Organisms and Manufactures: On the History of Plant Inventions', *Melbourne University Law Review*, 31: 539–568.

Pottage, A. and Sherman, B. (2010) *Figures of Invention: A History of Modern Patent Law*, Oxford: Oxford University Press.

Pottage, A. and Sherman, B. (2011) 'Kinds, clones, and manufactures', in M. Biagioli, P. Jaszi and M. Woodmansee (eds) *Making and Unmaking Intellectual Property*, Chicago: University of Chicago Press, pp 269–284.

Prado, J.R., Segers, G., Voelker, T., Carson, D., Dobert, R., Phillips, J. et al (2014) 'Genetically engineered crops: From idea to product', *Annual Review of Plant Biology*, 65(1): 769–790.

Praduroux, S. (2017) 'Objects of property rights: Old and new', in M. Graziadei and L.D. Smith (ed) *Comparative Property Law: Global Perspectives*. Cheltenham: Edward Elgar Publishing, pp 51–70.

Pratelli, A., Petri, M., Farina, A. and Lupi, M. (2018) 'Improving bicycle mobility in urban areas through its technologies: The SaveMyBike Project', in G. Sierpiński (ed) *Advanced Solutions of Transport Systems for Growing Mobility*, Cham: Springer International Publishing, pp 219–227.

Pratt, A.C. (2009) 'Cultural economy', in R. Kitchin and N. Thrift (eds) *International Encyclopedia of Human Geography*, Oxford: Elsevier, pp 407–410.

Proudhon, P.-J. (1994) *What is Property?*, Cambridge: Cambridge University Press.

Rabinow, P. (1996) *Making PCR: A Story of Biotechnology*, Chicago: University of Chicago Press.

Rabinow, P. (1997) 'Severing the ties: Fragmentation and dignity in Late Modernity', in P. Rabinow (ed) *Essays on the Anthropology of Reason*, Princeton, NJ: Princeton University Press, pp 129–152.

Radin, M.J. (1993) *Reinterpreting Property*, Chicago: University of Chicago Press.

Radin, M.J. (1996) *Contested Commodities*, Cambridge, MA: Harvard University Press.

Ragonnaud, G. (2013) *The EU Seed and Plant reproductive Material Market in Perspective: A Focus on Companies and Market Shares*, Policy Department B: Structural and Cohesion Policies.

Rakopoulos, T. (2022) 'Property as sovereignty in micro: The state/property nexus and the Cyprus Problem', *Journal of the Royal Anthropological Institute*, 28(3): 769–787.

Reece, J.D. and Haribabu, E. (2007) 'Genes to feed the world: The weakest link?', *Food Policy*, 32(4): 459–479.

Reitz, T., Sevignani, S. and van den Ecker, M. (2023) 'Eigentum im digitalen Kapitalismus. Ökonomie, Recht und Praxis', in T. Carstensen, S. Schaupp and S. Sevignani (eds) *Theorien des digitalen Kapitalismus: Arbeit, Ökonomie, Politik und Subjekt*, Berlin: Suhrkamp, pp 264–284.

Relly, E. (2023) 'Recursos genéticos e bioprospecção no Brasil: capitaloceno, protagonismo e os (des)caminhos até o Protocolo de Nagoya (2010)', *Caravelle. Cahiers du monde hispanique et luso-brésilien*, 119: 89–106.

Rheinberger, H.-J. (1995) *Kurze Geschichte der Molekularbiologie*, Berlin: Max Planck Institute for the History of Science.

Robé, J.-P. (2020) *Property, Power and Politics: Why We Need to Rethink the World Power System*, Bristol: Bristol University Press.

Rodriquez, J. (2006) 'Color-blind ideology and the cultural appropriation of hip-hop', *Journal of Contemporary Ethnography*, 35(6): 645–668.

Rose, C.M. (1985) 'Possession as the origin of property', *The University of Chicago Law Review*, 52(1): 73–88.

Rousseau, J.-J. (2011) *Discourse on the Origin and Foundations of Inequality among Men*, New York: Bedford/St. Martins.

Ryan, A. (1987) *Property*, Minneapolis: University of Minnesota Press.

Sánchez-Monge, E. (1993) 'Introduction', in M.D. Hayward, N.O. Bosemark, I. Romagosa and M. Cerezo (eds) *Plant Breeding Principles and Prospects*, Dodrecht: Springer Netherlands, pp 3–6.

Sanderson, J. (2017) *Plants, People and Practices: The Nature and History of the UPOV Convention*, Cambridge: Cambridge University Press.

Santesmases, M.J. (2013) 'The biological landscape of polyploidy: Chromosomes under glass in the 1950s', *History and Philosophy of the Life Sciences*, 35(1): 91–98.

Saraiva, T. (2016) *Fascist Pigs: Technoscientific Organisms and the History of Fascism*, Cambridge, MA: MIT Press.

Schaniel, W.C. and Neale, W.C. (1999) 'Quasi commodities in the First and Third Worlds', *Journal of Economic Issues*, 33(1): 95–115.

Schatzki, T.R., Knorr-Cetina, K. and von Savigny, E. (eds) (2001) *The Practice Turn in Contemporary Theory*, London: Routledge.

Scheffer, T. (2010) *Adversarial Case-Making: An Ethnography of English Crown Court Procedure*, Leiden: Brill.

Schiebinger, L.L. (2004) *Plants and Empire: Colonial Bioprospecting in the Atlantic World*, Cambridge, MA: Harvard University Press.

Schlager, E. and Ostrom, E. (1992) 'Property-rights regimes and natural resources: A conceptual analysis', *Land Economics*, 68(3): 249–262.

Scholz, A.H., Freitag, J., Lyal, C.H.C., Sara, R., Cepeda, M.L., Cancio, I. et al (2022) 'Multilateral benefit-sharing from digital sequence information will support both science and biodiversity conservation', *Nature Communications*, 13(1): 1086.

Schubert, J., Böschen, S. and Gill, B. (2011) 'Having or doing intellectual property rights? Transgenic seed on the edge between refeudalisation and napsterisation', *European Journal of Sociology*, 52(1): 1–17.

Schwartz, P.M. (2003) 'Property, privacy, and personal data', *Berkeley Law Scholarship Repository*, 117: 2056–2128.

Seeger, A. (2004) 'The selective protection of musical ideas: The "creators" and the dispossessed', in K. Verdery and C. Humphrey (eds) *Property in Question: Value Transformation in the Global Economy*, Oxford: Berg, pp 69–84.

Serres, M. (2007) *The Parasite*, Minneapolis: University of Minnesota Press.

Shapiro, C. (2000) 'Navigating the patent thicket: Cross licenses, patent pools, and standard setting', *Innovation Policy and the Economy*, 1: 119–150.

Sherman, B. (2008) 'Taxonomic property', *The Cambridge Law Journal*, 67(3): 560–584.

Sherman, B. and Bently, L. (2003) *The Making of Modern Intellectual Property Law: The British Experience, 1760–1911*, Cambridge: Cambridge University Press.

Siegrist, H. (2005) 'Kommentar: Eigentum und soziale Handlungsrechte im Übergang von der frühen Neuzeit zur Moderne. Die „Propertization" von Gesellschaft und Geschlecht', *Comparativ*, 15(4): 97–108.

Siegrist, H. (2006) 'Die Propertisierung von Gesellschaft und Kultur. Konstruktion und Institutionalisierung des Eigentums in der Moderne', *Comparativ*, 16(5/6): 9–52.

Singer, J.W. (2006) 'Nine-tenths of the law: Title, possession & sacred obligations', *Connecticut Law Review*, 38(4): 605–630.

Sivaraman, A. (2013) 'The Shield Act: A good attempt at curbing patent trolls that leaves us wanting more', *Journal of Business, Entrepreneurship and the Law*, 7(1): 209–240.

Slater, D. (2002) 'From calculation to alienation: Disentangling economic abstractions', *Economy and Society*, 31(2): 234–249.

Slaughter, S. and Rhoades, G. (2004) *Academic Capitalism and the New Economy: Markets, State, and Higher Education*. Baltimore: Johns Hopkins University Press.

Snell, K. (1939) 'Sortenschutz durch Registrierung', *Der Züchter*, 11(1): 22–24.

Soler, L., Zwart, S., Lynch, M. and Israel-Jost, V. (eds) (2014) *Science after the Practice Turn in the Philosophy, History, and Social Studies of Science*, New York: Routledge.

Sommer, F. (2022) *Äcker und Daten in Deutschland: Eine Ethnographie der Landrechte im Spiegel(kabinett) staatlicher Register*, PhD dissertation, Halle/Saale: Martin-Luther-Universität Halle-Wittenberg.

Star, S.L. (1989) 'The structure of ill-structured solutions: Boundary objects and heterogeneous distributed problem solving', in L. Gasser and M.N. Huhns (eds) *Distributed Artificial Intelligence*, San Francisco: Morgan Kaufmann, pp 37–54.

Star, S.L. (1990) 'Power, technology and the phenomenology of conventions: On being allergic to onions', *The Sociological Review*, 38(1_suppl): 26–56.

Star, S.L. (ed) (1995) *Ecologies of Knowledge: Work and Politics in Science and Technology*, Albany: State University of New York Press.

Star, S.L. and Griesemer, J.R. (1989) 'Institutional ecology, "translations" and boundary objects: Amateurs and professionals in Berkeley's Museum of Vertebrate Zoology, 1907–39', *Social Studies of Science*, 19(3): 387–420.

Stemerding, D., Beumer, K., Edelenbosch, R., Swart, J.A.A., de Vries, M.E., Ter Steeg, E. et al (2023) 'Responsible innovation in plant breeding: The case of hybrid potato breeding', *Plants*, 12(9): 1751.

Stengel, K., Taylor, J., Waterton, C. and Wynne, B. (2009) 'Plant sciences and the public good', *Science, Technology, & Human Values*, 34(3): 289–312.

Stengers, I. (1997) *Power and Invention: Situating Science*, Minneapolis: University of Minnesota Press.

Stengers, I. (2011) *Cosmopolitics II*, Minneapolis: University of Minnesota Press.

Stone, C.D. (2010) *Should Trees have Standing? Law, Morality, and the Environment*, 3rd ed, Oxford: Oxford University Press.

Stone, G.D. (2002) 'Both sides now: Fallacies in the genetic-modification wars, implications for developing countries, and anthropological perspectives', Current Anthropology, 43(4): 611–630.

Stone, G.D. and Glover, D. (2017) 'Disembedding grain: Golden rice, the green revolution, and heirloom seeds in the Philippines', *Agriculture and Human Values*, 34(1): 87–102.

Strathern, M. (1988) *The Gender of the Gift: Problems with Women and Problems with Society in Melanesia*, Berkeley: University of California Press.

Strathern, M. (1999) What is intellectual property after?, *The Sociological Review*, 47(1_suppl): 156–180.

Strathern, M. (2004) 'Losing (out on) intellectual resources', in A. Pottage and M. Mundy (eds) *Law, Anthropology, and the Constitution of the Social: Making Persons and Things*, Cambridge: Cambridge University Press, pp 201–233.

Strathern, M. (2009) 'Land: Intangible or tangible property?', in T. Chesters (ed) *Land Rights: The Oxford Amnesty Lectures 2005*, Oxford: Oxford University Press, pp 13–38.

Strathern, M. (2022) *Property, Substance, and Effect: Anthropological Essays on Persons and Things*, Chicago: HAU Books.

Swedberg, R. (2003a) *Principles of Economic Sociology*, Princeton, NJ: Princeton University Press.

Swedberg, R. (2003b) 'The case for an economic sociology of law', *Theory and Society*, 32(1): 1–37.

Tangens, R. (2005) 'Den Big Brother Award 2005 in der Kategorie "Wirtschaft" erhält die Saatgut-Treuhand Verwaltungs GmbH in Bonn vertreten durch ihren Geschäftsführer Dirk Otten', Bigbrotherawards.de, available from https://bigbrotherawards.de/2005/wirtschaft-saatgut-treuh and [Accessed: 13 December 2023].

Tellmann, U. (2022) 'The politics of assetization: From devices of calculation to devices of obligation', *Distinktion: Journal of Social Theory*, 23(1): 33–54.

Then, C. and Tippe, R. (2016), *Patente auf Pflanzen und Tiere: Jetzt müssen Europas Politiker handeln*, München: Kein Patent auf Saatgut!, available from: https://www.no-patents-on-seeds.org/sites/default/files/news/bericht_patente_auf_saatgut_zeit_zu_handeln_2016.pdf [Accessed: 10 December 2023].

Thirtle, C., Bottomley, P., Palladino, P., Schimmelpfennig, D. and Townsend, R. (1998) 'The rise and fall of public sector plant breeding in the United Kingdom: A causal chain model of basic and applied research and diffusion', *Agricultural Economics*, 19(1–2): 127–143.

Thomas, N. (1991) *Entangled Objects: Exchange, Material Culture, and Colonialism in the Pacific*, Cambridge, MA: Harvard University Press.

Thompson, D. (2014) *Circuits of Containment: Iron Collars, Incarceration and the Infrastructure of Slavery*, PhD dissertation, Ithaca, NY: Cornell University.

Timmermann, M. (2009) *Der Züchterblick: Erfahrung, Wissen und Entscheidung in der Getreidezüchtung*, Aachen: Shaker.

Tönnies, F. (1926) *Das Eigentum*, Wien: Braumüller.

Torma, F. (2017) 'Biofakte als historiografische Linsen: Die Technik- und Gesellschaftspolitik des Hybridmaises im geteilten Deutschland', *Technikgeschichte*, 84(2): 135–162.

Torma, F. (2018) 'Vom Retter zum Frankenstein. Soziotechnische Imaginationen zu "Mais"', in B. Gill, F. Torma and K. Zachmann (eds) *Mit Biofakten Leben: Sprache und Materialität von Pflanzen und Lebensmitteln*, Baden-Baden: Nomos, pp 27–58.

Torrisi, S., Gambardella, A., Giuri, P., Harhoff, D., Hoisl, K. and Mariani, M. (2016) 'Used, blocking and sleeping patents: Empirical evidence from a large-scale inventor survey', *Research Policy*, 45(7): 1374–1385.

Tsing, A. (2013) 'Sorting out commodities: How capitalist value is made through gifts', *HAU: Journal of Ethnographic Theory*, 3(1): 21–43.

Tsunewaki, K. (2003) 'Alien Cytoplasms', in B. Thomas (ed) *Encyclopedia of Applied Plant Sciences*, Oxford: Elsevier, pp 158–167.

Turner, B. (2017) 'The anthropology of property', in M. Graziadei and L.D. Smith (eds) *Comparative Property Law: Global Perspectives*, Cheltenham: Edward Elgar Publishing, pp 26–47.

Turner, B. and Wiber, M.G. (2022) 'Legal pluralism and science and technology studies: Exploring sources of the legal pluriverse', *Science, Technology, & Human Values*, 48(3): 457–474.

Uekötter, F. (2012) *Die Wahrheit ist auf dem Feld: Eine Wissensgeschichte der deutschen Landwirtschaft*, Göttingen: Vandenhoeck & Ruprecht.

Umphrey, M.M. (2011) 'Law in drag: Trials and legal performativity', *Columbia Journal of Gender and Law*, 21(2): 114–129.

van Dooren, T. (2007a) *Seeding Property: Nature, Human/Plant relations and the Production of Wealth*, PhD dissertation, Canberra: Australian National University.

van Dooren, T. (2007b) 'Terminated seed: Death, proprietary kinship and the production of (bio)wealth', *Science as Culture*, 16(1): 71–94.

van Dooren, T. (2008) 'Inventing seed: The nature(s) of intellectual property in plants', *Environment and Planning D: Society and Space*, 26(4): 676–697.

Van Overwalle, G. (2017a) 'Plant patents: From exclusivity to inclusivity', CVPO, [online] 1 March 2017, available from: https://cpvo.europa.eu/en/news-and-events/articles/plant-patents-exclusivity-inclusivity [Accessed: 11 December 2023].

Van Overwalle, G. (2017b) 'Patent pools and clearinghouses in the life sciences: Back to the future', in D. Matthews and H. Zech (eds) *Research Handbook on Intellectual Property and the Life Sciences*, Cheltenham: Edward Elgar, pp 304–333.

Vasile, M. (2007) 'The sense of property, deprivation and memory in the case of Obştea Vrânceană', *Sociologie Românească*, 2: 114–129.

Verdery, K. and Humphrey, C. (2004) 'Introduction: Raising questions about property', in K. Verdery and C. Humphrey (eds) *Property in Question: Value Transformation in the Global Economy*, Oxford: Berg, pp 1–25.

Verriet, J. (2013) 'Ready meals and cultural values in the Netherlands, 1950–1970', *Food and History*, 11(1): 123–153.

Vinsel, L. (2021) 'You're doing it wrong: Notes on criticism and technology hype', *Medium*, [online] 1 February 2021, available from: https://sts-news.medium.com/youre-doing-it-wrong-notes-on-criticism-and-technology-hype-18b08b4307e5 [Accessed 11 December 2023].

Voicu, S. and Vasile, M. (2022) 'Grabbing the commons: Forest rights, capital and legal struggle in the Carpathian Mountains', *Political Geography*, 98: 1–12.

von Benda-Beckmann, F. (1979) *Property in Social Continuity*, Dordrecht: Springer Netherlands.

von Benda-Beckmann, F. (1995) 'Anthropological approaches to property law and economics', *European Journal of Law and Economics*, 2(4): 309–336.

von Benda-Beckmann, F., von Benda-Beckmann, K. and Wiber, M. (eds) (2006a) *Changing Properties of Property*, New York: Berghahn Books.

von Benda-Beckmann, F., von Benda-Beckmann, K. and Wiber, M. (2006b) 'The properties of property', in F. von Benda-Beckmann, K. von Benda-Beckmann and M. Wiber (eds) *Changing Properties of Property*, New York: Berghahn Books, pp 1–39.

Waltenberger, S. (2020) *Deutschlands Ölfelder: Eine Stoffgeschichte der Kulturpflanze Raps (1897–2017)*, Paderborn: Ferdinand Schöningh, Brill.

Watson, J.D. and Crick, F.H.C. (1953) 'Molecular structure of nucleic acids: A structure for deoxyribose nucleic acid', *Nature*, 171(4356): 737–738.

Weber, M. (1950) *General Economic History*, Glenncoe, IL: Free Press.

Weber, M. (1978) *Economy and Society: An Outline of Interpretive Sociology*, Berkeley: University of California Press.

Weber, M. (2008) *Roman Agrarian History: In Its Relation to Roman Public & Civil Law*, Claremont, CA: Regina Books.

Wesche, T. (2023) *Die Rechte der Natur: Vom nachhaltigen Eigentum*, Berlin: Suhrkamp.

Whipple, T. (2023) 'I have owned 11 bikes. This is how they were stolen', *The Times*, [online] 27 November, available from: https://www.thetimes. co.uk/article/i-have-owned-11-bikes-this-is-how-they-were-stolen-d3r553gx3 [Accessed 27 November 2023].

Wiber, M.G. (2015) 'Property as boundary object: Normative versus analytical meanings', *The Journal of Legal Pluralism and Unofficial Law*, 47(3): 438–455.

Wieland, T. (2004) *'Wir beherrschen den pflanzlichen Organismus besser, …'; wissenschaftliche Pflanzenzüchtung in Deutschland, 1889–1945*, München: Deutsches Museum.

Wieland, T. (2006) 'Scientific theory and agricultural practice: Plant breeding in Germany from the late 19th to the early 20th century', *Journal of the History of Biology*, 39(2): 309–343.

Wieland, T. (2009) 'Autarky and Lebensraum: The political agenda of academic plant breeding in Nazi Germany', *Journal of History of Science and Technology*, 3: 14–34.

Wieland, T. (2011) 'Von springenden Genen und lachsroten Petunien. Epistemische, soziale und politische Aspekte der gentechnischen Transformation der Pflanzenzüchtung', *Technikgeschichte*, 78(3): 255–278.

Windpassinger, W. (1998) '"Die Gebühren für Saatgut hat den Bauern allein der Staat eingebrockt"', *Passauer Neue Presse*, 25 August 1998.

Winickoff, D., Jasanoff, S., Busch, L., Grove-White, R. and Wynne, B. (2005) Adjudicating the GM food wars: Science, risk, and democracy in world trade law, *Yale Journal of International Law*, 30(1): 82–123.

Woolf, V. (1929) *A Room of One's Own*, London: Hogarth.

Würtenberger, G. (2014) 'Nachbauvergütungen: Eine kritische Bestandsaufnahme', in A. Metzger (ed) *Rechtsschutz von Pflanzenzüchtungen. Eine kritische Bestandsaufnahme des Sorten-, Patent- und Saatgutrechts*, Tübingen: Mohr Siebeck, pp 105–114.

Wynne, B. (2001) 'Creating public alienation: Expert cultures of risk and ethics on GMOs', *Science as Culture*, 10(4): 445–481.

Young, J.O. (2008) *Cultural Appropriation and the Arts*, Malden, MA: Blackwell.

Index

References to endnotes show both the page number and the note number (231n3).